Title Page

Focused Operations Performance Improvement

Brice Alvord, MBA

Brice Alvord

Limits of Liability Disclaimer of Warranty

ALERA Publishing Group
PO Box 6111
Wyomissing, PA 19610

Copyright

Published by ALERA Publishing Group, Inc. Wyomissing, PA 19610

Printed in the United States of America

ISBN 978-1-300-68698-9

About This Manual

Introduction

This manual is a training workbook for helping participants to understand and apply the fundamentals of improving operations performance. It is designed to provide clear and simple explanations of concepts, procedures and techniques that make up this program.

How To Use This Manual

There are four ways to find information in this manual:
1. The table of contents, which lists the sections of the manual
2. The table of illustrations, which lists the illustrations found in this manual
3. The map title at the top of every page which provides a quick reference as to which section the manual is open to

Description

The sections described in this manual include:
- Fundamentals of operations performance analysis
- Defining the current situation
- Factor analysis
- Establishing optimal conditions
- Determining causal factors
- Creating a reliable methodology -CEDAC

In This Manual

The sections described in this manual are located as indicated below:

Focused Operations Performance Improvement

Table of Contents

Table of Figures

Continued on next page

Table of Figures, Continued

Table of Tables

Fundamentals of Operations Performance Analysis

Overview

Introduction

Typically Operations Performance Improvement (OPI) activities yield less than anticipated results for a variety of reasons. The most common reasons often include:

- OPI team members have inadequate knowledge of equipment structures and operating principles which results in an incomplete analysis
- The abnormal occurrence is not properly defined, resulting in improper counter measures being applied.

Goal

"Focused Operations Performance Improvement" shows your OPI team how to develop effective solutions to persistent performance problems. Your team will learn how to isolate and understand the root cause of defects and failures within equipment mechanisms and peripheral systems. They will learn how to apply a systematic approach for effectively controlling those causes.

In This Chapter

The topics covered in this chapter include:

Topic	See Page
Overall Equipment Effectiveness	2
Losses	13
Conventional problem solving techniques	27
Problem analysis approach	45

OVERALL EQUIPMENT EFFECTIVENESS

OEE Overview

Introduction	Overall Equipment Effectiveness (OEE) is a universal measurement that has been used worldwide for over 10 years. It is a formula to measure the efficiency of production line equipment. In short, OEE measures the ratio of first-pass acceptable product actually produced to the theoretical amount that could be produced under optimal conditions.
Why Overall Equipment Effectiveness	Overall equipment effectiveness is a measurement used to indicate how effectively machines are running.
In This Section	The topics described in this section are located as indicated below:

Topic	See Page
Losses reduce overall equipment effectiveness	5
Availability	9
Performance	11
Quality	12

Continued on next page

OEE Overview, Continued

OEE vs. Efficiency	What do we mean by overall equipment effectiveness? Many people are familiar with the idea of "efficiency," which usually reflects the quantity of parts a machine or a person can produce in a certain time. OEE is different from efficiency in several ways
Quantity Over Time Is Only Part of OEE	A machine's overall effectiveness includes more than the quantity of product it can produce in a shift. When we measure overall equipment effectiveness, we account for efficiency as one factor: • Performance In addition to performance, however, OEE includes two other factors: • Availability • Quality
Performance	A comparison of the actual output with what the machine should be producing in the same time.
Availability	A comparison of the potential operating time and the time in which the machine is actually making products
Quality	A comparison of the number of products made and the number of products that meet the customer's specifications
OEE Gives A Complete Picture	When you multiply performance, availability, and quality, you get the overall equipment effectiveness, which is expressed as a percentage. OEE gives a complete picture of the machine's "health"-not just how fast it can make parts, but how much the potential output was limited due to lost availability or poor performance. On subsequent pages, we will look more closely at these three elements and how they work together.

Continued on next page

OEE Overview, Continued

Effectiveness Focuses on the Equipment, Not the Person	Unlike some uses of the efficiency measure, OEE monitors the machine or process that adds the value, not the operator's productivity. When we measure OEE, we look at how well the equipment or process is working.
The Purpose of Measurement Is Improvement	Unlike some uses of the efficiency measure, OEE monitors the machine or process that adds the value, not the operator's productivity. When we measure OEE, we look at how well the equipment or process is working.
Improving Equipment Processes	Measuring OEE is not an approach for criticizing people. It is strictly about improving the equipment or process. Used as an impartial daily snapshot of equipment conditions, OEE promotes openness in information sharing and a no-blame approach in handling equipment-related issues.

These key differences highlight the importance of OEE as a balanced measure that helps support improvement and profitability. |

Losses Reduce Overall Equipment Effectiveness

Introduction

What makes machines less effective than they could be? The ideal, totally effective machine could run all the time (or whenever needed). It could maintain its maximum or standard speed all the time. It would never make defective products.

But most machines aren't ideal. They cannot run continuously. They cannot maintain maximum speed without problems...and they make defects.

Equipment-Related Losses

These problems are familiar forms of waste – they don't add value to the products. They reduce a machine's effectiveness, as measured by the OEE. The conditions that cause these machine problems are called equipment-related losses. Understanding the different types of equipment-related losses will give you a framework for applying OEE and participating in improvement activities to reduce the losses.

The equipment-related losses that are important for OEE are linked to the three basic elements measured in OEE: availability, performance, and quality.

Six Major Losses

There are six major losses that fall into the following three categories:
- Availability
- Performance:
- Quality:

Availability

Downtime losses include:
- Failures
- Setup time

Performance

Speed losses include:
- Minor stoppages
- Reduced operating speed

Quality

Defect losses include:
- Scrap and rework
- Startup loss

Continued on next page

OEE Overview, Continued

Basic Framework   Although some companies link individual losses to different OEE categories, or add other losses that are especially significant for their operations, this basic framework is a useful starting point for many companies. Figure 1 below gives a visual image of the way in which these losses reduce the overall equipment effectiveness of a machine.

Figure 1: Impact of Losses on OEE

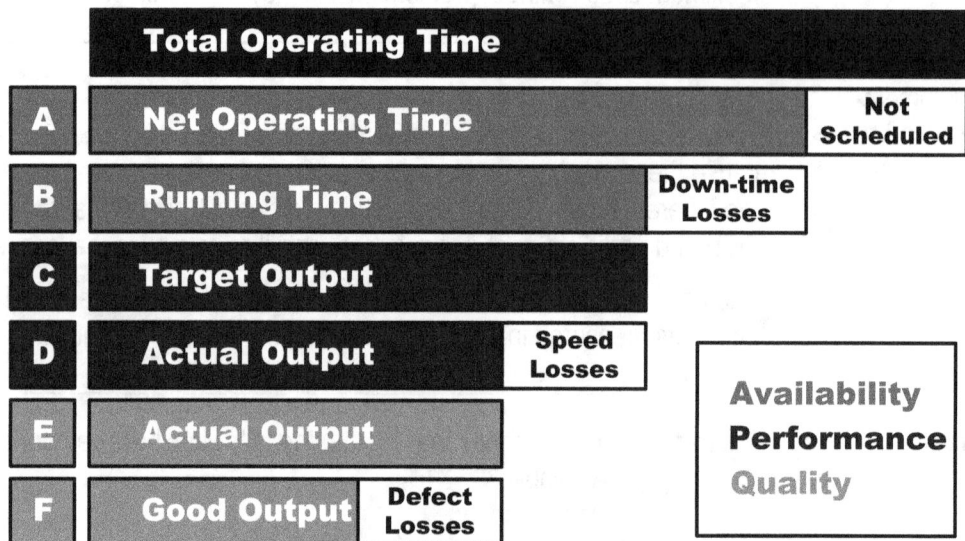

OEE = B/A X D/C X F/E X 100

Visualizing OEE and the Losses   Figure 1 makes it easy to see how OEE is derived from the three elements, expressed as fractions. Each pair of bars stands for one of the fractions--availability (B/A), performance (D/C), and quality (F/E). The fractions are often multiplied by 100 to turn them into percentages or rates.

Continued on next page

OEE Overview, Continued

Availability	Bars A and B represent availability. Unscheduled time shortens the total operating time,* leaving net operating time (A). But the machine is frequently down during some of that time, usually due to breakdowns and setup. Subtracting that downtime leaves the running time (B) in which the machine is making product.
Example	$$\frac{\text{Running Time}}{\text{Net Operating Time}} = \frac{300 \text{ minutes}}{400 \text{ minutes}} = .75 \text{ availability x } 100 = 75\%$$
Performance	Bars C and D represent performance. During the running time, the machine could produce a target output quantity (C) if it ran at its designed speed the whole time. But losses such as minor stoppages and reduced operating speed lower the actual output (D)
Example	$$\frac{\text{Actual Output}}{\text{Target Output}} = \frac{12,000 \text{ Units}}{20,000 \text{ Units}} = .60 \text{ Performance (x } 100 = 60\%)$$
Quality	Bars E and F represent quality. Of the actual output (E), most of the product is good output (F). But usually some output falls short of the specified quality and must be scrapped or reworked. Scrap is often produced during machine startup as well, lowering the yield from the materials.
Example	$$\frac{\text{Good Output}}{\text{Actual Output}} = \frac{11,760 \text{ Units}}{12,000 \text{ Units}} = .98 \text{ Quality (x } 100 = 98\%)$$
How Losses Impact Production	Figure 1 shows how losses to availability, performance, and quality compound to reduce the amount of good output a machine can produce during a shift. You can improve quality to raise the quantity of good output a little bit--but the total quantity won't rise dramatically unless you also improve both performance and availability.

Continued on next page

OEE Overview, Continued

OEE Formula	The formula at the bottom of Figure 1 shows how to multiply the three elements to get the OEE.

Example	.75 {Availability)	X	.60 (Performance)	X	.98 (Quality)	X 100	= 44% OEE

Availability:

Downtime Losses	There are basically two types of downtime losses that we are concerned with: • Failures • Setup Time
Failures	Availability is reduced by equipment failures, which are a common occurrence in many plants. Machines used for production generally have lots of moving parts and various subsystems in which things can go wrong. When they do, the machine breaks down and stays down until repairs are completed. Many of the causes of machine failure give warning signs before the machine actually breaks. In Chapter 4 we will look at how autonomous maintenance activities can help spot early trouble signs in time to prevent major breakdowns.
Setup Time	Availability is also reduced by the time it takes to set up the machine for a different product. In addition to changing the value adding parts, a changeover requires some preparation or make ready. It may involve cleaning and making adjustments to the machine to get stable quality in the next product. Too often, it also involves running around to find tools, parts, or people.

Continued on next page

OEE Overview, Continued

Other Losses to Availability

Failures and setup losses were the original losses counted as downtime that reduces availability. Some companies also track other losses as downtime, depending on what losses they are trying to improve. Cutting tool loss, startup loss, and time not scheduled for production are three other losses tracked as downtime at some plants.

Startup Loss

Startup loss is traditionally included as a defect loss, since its essence is the production of defective products during startup. However, startup loss involves lost time until good production can be stabilized, so it is logical to subtract it from available time as well.

Time Not Scheduled for Production

In some companies, when machines are stopped for meetings, preventive maintenance, or breaks, the time is considered "not scheduled" and is not counted in the availability rate (see Figure 2-5). Other companies recognize that even necessary activities like these reduce the available production time. They may decide to consider time "not scheduled" as a downtime loss that lowers the availability rate.

Performance

Speed Losses	Throughput loss due to lower than Control Point speed operation, including: • Reduced Operating Speed • Minor Stoppages
Reduced Operating Speed	Machines often run at speeds slower than they were designed to run. One reason for slower operation is unstable product quality' at the designed speed. In other cases, people don't realize that the equipment is designed to run faster. We will look in Chapter 3 at how to determine speed for the OEE calculation.
Minor Stoppages	Minor stoppages are events that interrupt the production flow without actually making the machine fail. They often occur on automated lines, for example when product components snag on the conveyor (see Figure 2-6). Minor stoppages can make it impossible to run automated equipment without someone to monitor it. These stoppages may seem like petty annoyances, but they add up to big losses at many plants. Minor stoppages last only a few seconds, so we don't try to log the time lost. Instead, we include them in performance losses that reduce the product output.

Quality

Defect Losses	There are two types of loss associated with quality: • Scrap and Rework • Startup Loss
Scrap and Rework	Products that do not meet customer specifications are a familiar loss. Clearly, scrap that cannot be reused is a waste of materials. Even when products can be reworked, the effort spent to process them twice is a waste.
Startup Loss	Many machines take time to reach the right operating conditions at startup. In the meantime, they may turn out defective products while operators test for stable output. Some companies simply include this startup loss in scrap and rework; others single it out as a specific loss to track. Quality problems happen when the optimum conditions do not exist at the moment when a person or machine works on the product.

IDENTIFYING LOSSES

Introduction	This section identifies the major equipment-related losses that lower equipment efficiencies and performance

6 Big Losses	Equipment problems reduce productivity in six ways, often referred to as the six big losses. The following is a list of the "six big losses" and the reasons why they lower the overall efficiency of a plant:

Loss	Reason
Breakdowns	Breakdowns result in lost time that the equipment could be producing good product.
Setup and adjustment loss	Changing tools and waiting for the next good part to be produced produces lost time that good product could be produced.
Idling and minor stoppages	Minor stoppages result is lost time because the equipment is not producing good product.
Reduced speed	The number of defects increases when equipment is not run at its specified speed.
Defects and rework	The number of defects increases when equipment is improperly setup.
Startup and yield loss	Startup results in lost time because the equipment cannot produce good product until fully warmed up.

In This Section	This section contains the following topics:

Breakdowns

Introduction	When one thinks about the causes of breakdowns, it's clear that equipment doesn't breakdown on its own. Sometimes a single abnormality may cause the breakdown, but many breakdowns result from the cumulative effects of several abnormalities. For this reason, dealing with breakdowns is more than a matter of fixing problems as they are noticed.

Contributing Factors

Breakdowns arise from any number of abnormalities in the equipment. Some problems are unavoidable, but almost all breakdowns are related to something that could have been prevented:

- Contamination
- Lack of lubrication
- Loose nuts or bolts
- Neglect of worn parts
- Mistakes in operation, changeover or repairs

Figure 2: Breakdowns According to Cause

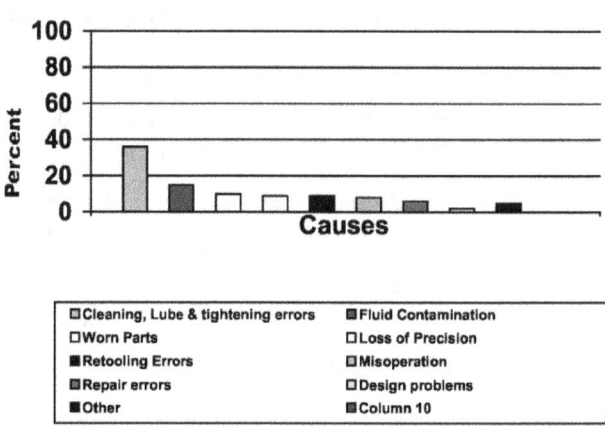

Continued on next page

Breakdowns, Continued

Two Types of Breakdowns	Two types of breakdowns can occur when a piece of equipment loses a given function or functions: • Function-loss breakdown: The complete loss of equipment function. Also known as sporadic breakdown • Function-reduction breakdown: This kind of breakdown occurs when a piece of equipment suffers from some partial loss of function. The machine still runs but experiences losses such as defects, minor stoppages, and reduced speed.
Attitude Causes Failure	Most often, equipment doesn't just fail; the failure is generally caused by people through neglect of procedures. Breakdowns also occur as the result of the wrong attitude, i.e. "breakdowns are inevitable" or "That is not my responsibility" or "That is the job of the maintenance department!"
Design and Function of Components	Production equipment is complex; it is made up of a large number of gears, switches, connectors, screws, bolts, shafts, motors, cylinders, sensors, and other parts. The equipment will operate normally as long as each of these components performs according to design and function. If the equipment is operated improperly, or poorly maintained, the function of the components will deteriorate and lead to failure and cause a breakdown.

Continued on next page

Breakdowns, Continued

Figure 3: The Bathtub Curve

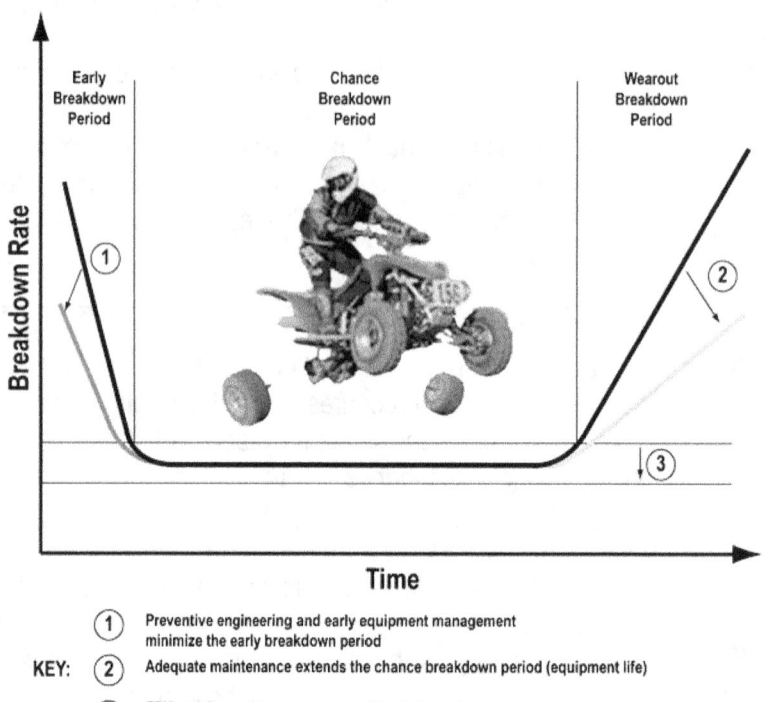

KEY:
① Preventive engineering and early equipment management minimize the early breakdown period
② Adequate maintenance extends the chance breakdown period (equipment life)
③ TPM activites reduce occurrence of breakdowns to zero

Table 1: When Breakdowns Occur

Breakdowns fall into three categories as shown in Figure 3 are based on when they occur in the life of the equipment, these categories are explained in the following table

Category	Explanation
1. Early Breakdown	Breakdowns occur in the start-up and initial use of equipment due to errors in design, fabrication, or installation
2. Chance Breakdown	Breakdowns in the chance category occur because of unforeseen reasons
3. Wear Out Breakdown	Wear Out related breakdowns occur because of wear or tear and deterioration as the equipment ages.

Continued on next page

Breakdowns, Continued

Types of Deterioration

Category 3 of Table 1 points out that deterioration is the cause of this type of breakdown. There are two types of deterioration found in equipment:
- Natural
- Accelerated

Natural deterioration is performance loss caused by physical deterioration that occurs over time when the equipment is operated and maintained properly.

Accelerated deterioration is caused by people neglecting to follow procedures and manufacturer's recommendations. Failure to establish or follow PM procedures is one of the main causes of this type of deterioration.

Setup and Adjustment Loss

Introduction	Setup and adjustment losses refer to the time it takes to change tools and dies as well as the time required to get the equipment running at full production. In other words, it is the amount of time lost until the next good part is produced at normal running speed.
Impact of Diversity of Products	One of the major consideration in controlling set up and adjustment loss is the number of products being produced on a particular machine. It is not uncommon for companies with a rather diverse list of products to find that the set up and adjustment takes more time that the production of product.
Types of Setup Operations	The key to reducing setup and adjustment loss is to reduce set up times and simplify the process. These reductions require that the Reliability Team understand set-up operation. There are two types of set up operations: • Internal • External
Internal Setup	This type of set up is performed only when the equipment is shut down. Internal set up operations hinder set up time reductions.
External Setup	This type of set up can be performed while the equipment is operating. This can help reduce set up time by having spare equipment parts that can be set up while the previous lot is being run.
Reducing Setup and Adjustment Times	The Reliability Team can significantly reduce startup and adjustment losses by implementing a quick changeover program. This process also know as SMED is typically implemented in three stages: 1. Separating internal and external set up 2. converting internal setup to external setup 3. streamlining all aspects of the set up operation

Idling and Minor Stoppages

Introduction

Minor stoppages are different from normal breakdowns. They are situations when the equipment either shuts down or idles because of a temporary problem

Lead to Major Losses

Minor stoppages have many negative effects on production, which include:
- Reducing the performance rate of equipment while minor stoppages and delays are corrected
- Delaying dependent equipment while stoppages are corrected
- Creating worn or deformed products while they are stuck or held up during a minor stoppage
- Wasting electricity and fuel by shutting down or idling machines.

Not Taken Seriously

Companies often don't take minor stoppages seriously even though they can lead to huge losses. The following reasons for this include:
- The size of the loss isn't obvious
- The symptoms are treated and not the real problem
- Not enough on-site inspections and observations are performed.

Reduced Speed

Introduction	A speed loss is an hourly loss that occurs when equipment is run at a lower speed than its standard or rated speed. Speed loss is not the amount of parts produced per hour.
Cause of Speed Loss	Speed losses are caused by reduced operating speed. When the equipment is run at higher operating speeds, quality defects and minor stoppages frequently occur. The equipment is therefore required to operate at lower speeds where these defects and stoppages do not occur. Speed loss is measured in terms of the ration of theoretical to actual operating speed.
Reasons for Occurrence	Before creating countermeasures for speed loss it is important to understand the reasons occur. The following are three bad beliefs that cause speed loss • The current machine cycle time is normal • The machine cycle must be slowed down to prevent defects • The equipment was poorly designed

Yield Loss

Introduction	One of the most overlooked losses occurs during the start up of a line after a shift change, start of a new week, or after an equipment changeover for a new product run.
Causes	Yield losses are caused by unused or wasted raw materials leading to : • Scraps • Rejects • Chips.
Categories of Loss	These losses are categorized into two groups: • Raw material losses resulting from product designs, manufacturing methods, and equipment restrictions • Adjustment losses resulting from quality defects associated with stabilizing operating conditions.
Determining Yield Loss	Yield losses are determined by adding the set and adjustment losses plus yield losses in terms of time <u>AND</u> material loss.

Defects and Rework Losses

Introduction

These losses are caused by off-specification or defective products manufactured during normal operation. These products must be reworked or scrapped.

Determining Rework Losses

Defects and rework losses consist of the labor required for rework of the product and the cost of the materials to be scrapped. The magnitude of these losses is measured by the ratio of quality products to total production.

Chronic Losses vs. Sporadic Losses

Introduction	Equipment failures and defects appear in two ways: as sporadic or chronic tosses. Sporadic Losses indicate sudden, often large deviations from the norm (current performance and quality levels). Chronic tosses, on the other hand, indicate smaller, frequent deviations that gradually have been accepted as normal
Sporadic Losses Are Easy to Correct	Sporadic losses, as the name implies, occur suddenly and infrequently. Typically, they result from a single cause that is relatively easy to identify. Also, because cause-and-effect relationships in sporadic losses are fairly dear, corrective measures are usually easy to formulate.
	For example, quality defects may arise when a jig has become abraded to a point where it no longer supports the required precision. Or a spindle may suddenly vibrate excessively, causing unacceptable dimensional variations in the product.
	Such problems are usually resolved through measures that restore the process condition or component to its immediately previous state.
Chronic Losses Have Obscure Causes	Chronic losses, on the other hand, live up to their name by resisting a wide variety of corrective measures. They require innovative, "breakthrough" measures that restore the mechanism or component to its *original,* defect-free state.
	Unlike sporadic losses, chronic losses are the products of complex, tangled cause-and-effect relationships. Tracking down their causes can be arduous. The reason is simple—chronic tosses rarely have just one cause, so it is difficult both to identify causes and to clarify their effects. This makes it equally difficult to devise effective countermeasures
	In nearly every case, while countermeasures may bring temporary improvement, the situation gets worse again with time. Eliminating such losses completely is a major challenge—one that conventional approaches can never overcome. What we need are new conceptual toots

Continued on next page

Chronic Losses vs. Sporadic Losses, Continued

Chronic Losses Have Obscure Causes

Chronic losses, on the other hand, live up to their name by resisting a wide variety of corrective measures. They require innovative, "breakthrough" measures that restore the mechanism or component to its *original,* defect-free state.

Unlike sporadic losses, chronic losses are the products of complex, tangled cause-and-effect relationships. Tracking down their causes can be arduous. The reason is simple—chronic losses rarely have just one cause, so it is difficult both to identify causes and to clarify their effects. This makes it equally difficult to devise effective countermeasures

In nearly every case, while countermeasures may bring temporary improvement, the situation gets worse again with time. Eliminating such losses completely is a major challenge—one that conventional approaches can never overcome. What we need are new conceptual tools.

Understanding the Nature of Chronic Loss

Defects and equipment failures persist on the shop floor because people try to tackle chronic tosses without understanding what this entails. Grasping the nature of chronic tosses is an important prerequisite to eliminating them through improvement.

Broadly speaking, there are two types of chronic loss
- The problem is produced by a single cause, but the cause varies from one occurrence to the next.
- The problem is produced by a combination of causes, which also varies from one occurrence to the next

Continued on next page

Chronic Losses vs. Sporadic Losses, Continued

Single Cause   Suppose for a given problem there are ten potential causes, A through J. Each time the problem occurs, the cause is different. Sometimes it may be A, sometimes C, or D, and so on. Consequently, measures focused on only one specific cause (A, for example) cannot control the problem.

For example, the finishing process which hones inner and outer races in ball bearing housings is critical to the quality of bearings. Consider a situation in which 1 to 2 percent of defects involve roughness in the races. Conditions that could conceivably produce such roughness include a misshapen, poorly mounted, or clogged grindstone; improper dressing; a loose grindstone holder or spindle; and imprecise curve grinding.

Possible causes range from things done in the previous process (such as curve grinding) to actions during the setup procedure (such as attaching the grinder). The actual cause of the defect, however, may change from one instance to the next.

To solve a problem like this the improvement team should examine *all* possible factors and restore them to their original conditions if necessary, while ensuring that correct dimensions and configurations are maintained. These actions are required because while there is no problem when causes can be identified, in practice, identifying them all is often very difficult.

Continued on next page

Chronic Losses vs. Sporadic Losses, Continued

Combination of Causes

In some cases a combination of multiple and overlapping causes generates the problem. To make matters worse, each time the problem occurs, a different combination of factors may be involved. Today it may be factors A, B, and C; tomorrow A, C, G, and H.

For example, in a polishing process using an internal grinding machine, out-of-roundness defects are sometimes generated by overlapping factors. These can include dimensional variations in the raw material, a worn reference plate on the work piece mounting surface, vibration in the grindstone spindle, and insufficient quill stiffness. Since all these factors may contribute in combination to any single occurrence of the defect, the team should consider corrective action for all of them, resisting the temptation to target just one.

Failing to appreciate how chronic losses occur is a major obstacle to their elimination. We tend to focus too narrowly on a given cause because we have not sufficiently understood the *phenomenon:* the physical event or precisely what happens to produce the defect in question. Avoiding this common pitfall is essential. Even when the measures taken against a single targeted cause are effective, the improvement is often temporary. The problem will ultimately resurface, because we have failed to eliminate other causes.

Problems in Reducing Chronic Loss

To achieve a lasting reduction in chronic losses we must do three things:
- Identify all factors that conceivably contribute to a loss.
- Thoroughly investigate each factor.
- Eliminate any malfunctions or suboptimal conditions discovered in the process.

Other Reasons

Chronic losses persist for another reason. Even when people understand the nature of chronic loss, they continue to use a flawed problem-solving approach. Three types of errors are common in chronic loss analysis:
- Phenomena are insufficiently stratified and analyzed.
- Some factors related to individual phenomena are overlooked.
- Abnormalities hidden in individual factors are not addressed.

CONVENTIONAL PROBLEM SOLVING TECHNIQUES

Introduction	A problem solving model is conceptual framework for addressing problems, not a formula for solving them.
	People are the actual problem solvers. But teams of people usually need an agreed-upon framework to keep them focused on task.
	Though problem-solving models can be highly sophisticated and technical, the model covered in this section has just five simple steps. Despite its simplicity, it is comprehensive enough to address all but the most technical problems. But because of its simplicity, your team is likely to remember and use it.
Purpose of a Model	A problem-solving model can help a team stay on track and work efficiently. Without a model the team may overlook many of its options and step into still other problems.
	With a model, a team has a framework to help:
	• Air everyone's concerns Look beyond
	• Explore all solutions
	• Anticipate problems
	• Follow through
	• Work productively
In This section	The topics covered in this section include:

Using the Problem Solving Model

Introduction

The sequence begins with step "1"- Identify the problem" and proceeds clockwise through all six steps:
- Identify the problem
- Analyze The Problem
- Evaluate Alternatives
- Test Implement The Solution
- Standardize The Solution

Figure 4: 5 Step Problem Solving Model

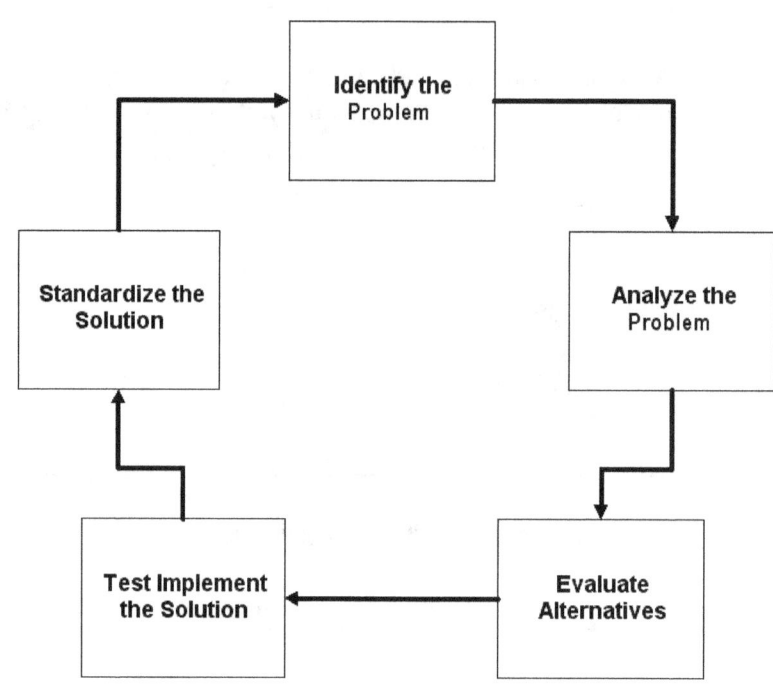

Continued on next page

Using the Problem Solving Model, Continued

Model Characteristics	The steps shown in Figure 4 are arranged in a circle to emphasize the cyclic continuous nature of the problem-solving process. The model has several important characteristics: • There are no branches or choice points. All five steps are required in the order shown. When one step is completed, your team proceeds counter clockwise to the next step. • The steps are repeatable. At any step, your team may decide to return to and repeat an earlier step. Analyzing a problem, for instance can lead back to re-identifying it. • The process is continuous. Implementing a solution does not end the process. • Though the steps have discrete names, there is no clear demarcation between them. Identifying and analyzing a problem frequently overlap.
1st Step	The first step – identify the problem is a broad review of the current situation. In most environments, problems, or improvement opportunities are easily identified. But choosing just any problem may not reap the benefits your efforts deserve. To get the most out of the time that will be invested completing the problem solving steps, focus on a customer-related problem When identifying the problem make sure that you: • Show the need for improvement in measurable terms • State the problem • Establish an interim target and a date for achieving this improvement
Useful tools for Identifying	Useful tools for identifying a problem include: • Brainstorming • Interview • Survey • List reduction • Matrix

Continued on next page

Using the Problem Solving Model, Continued

Guidelines For Developing A Problem Statement

When developing a problem statement:
- Be specific
- Describe a problem, not a symptom
- Relate the current situation to what is desired
- Be free of causes and solution

Guidelines for Setting Targets

When setting targets you must remember to:
- Express targets quantitatively
- Be aggressive in your selection

Analyze the Problem

When analyzing the problem:
- Identify the root cause(s) of the problem
- Verify each root cause
- Identify the root causes most responsible for the problem
- Targets should be changed as the situation changes
- Establish a long-term target and define intermediate targets

Useful Tools for Analyzing

Useful tools for analyzing a problem include:
- Cause-and-effect diagram
- Flowchart
- Pareto chart
- Brainstorming
- Check sheet

Continued on next page

Using the Problem Solving Model, Continued

Guidelines For Collecting Data	When collecting data: • Establish a purpose before collecting data • Determine if the indicators are reliable • Track all data needed • Record the data carefully
Evaluate Alternatives	When evaluating alternatives: • Identify actions that will reduce or eliminate the root cause(s) • Determine which actions will lead to the targeted level of improvement • Plan the implementation of selected solutions
Useful Tools for Evaluating	Useful tools for evaluating alternatives include: • Brainstorming • Interview • Survey
Guidelines for Developing Alternative Solutions	When developing alternative solutions: • Be creative -- identify as many potential actions as possible • Don't be constrained by current practice • Be supportive

Continued on next page

Focused Operations Performance Improvement

Using the Problem Solving Model, Continued

Factors to Consider When Choosing Solutions

Factors to consider when choosing solutions include:
- Effectiveness
  - **Has this been tried before?**
  - **Will it solve all or part of the problem?**
  - **Will it achieve the target for improvement?**
- Feasibility
  - Can we implement this solution?
  - Is it practical?
- Timeliness
  - How fast will it work?
  - Is it a long – or short-term solution?
  - Can we afford to wait?
- Customer-oriented
  - Does it satisfy identified customer requirements?
  - Will it improve service quality?

What to Consider

Consider the following:
- Manpower (people)
  - Whose support will be needed to successfully implement the solution?
- Materials
  - Will your solutions require that new or different materials be utilized?
  - Who will procure them?
- Methods
  - How will those involved learn how to implement what you propose?
  - How will you know if your solutions are working?
- Machinery
  - Will your solutions require that new or different equipment be utilized?

Elements of Planning

Elements of planning include:
- The objective is clearly stated
- Each activity is defined
- Responsibility assigned
- Due dates are established

Continued on next page

Using the Problem Solving Model, Continued

Test Implement the Solution

When test implementing the solution:
- Implement the plan
- Help the solutions succeed
- Show measurable improvement
- If measurable improvement is not evident, restate the problem

Useful Tools for Test Implementing

Useful tools for test implementing the solution include:
- Line graph
- Pareto chart
- Pie chart
- Bar chart
- Histogram
- Check sheet

Ensuring Effective Solutions

In order to ensure effective solutions, you should:
- Communicate the plan
- Monitor plan implementation
- Reinforce each other
- Adjust when necessary

Continued on next page

Using the Problem Solving Model, Continued

Standardize the Solution

When standardizing the solution, remember to:
- Ensure that your solutions are made permanent
- Determine if the solutions will be effective elsewhere

Useful Tools for Standardizing

Useful tools for standardizing the solution include:
- Flowcharting
- Brainstorming

Steps for Maintaining the Gains

To maintain your gains:
- Make periodic checks
- Clarify work activities
- Develop and follow procedures
- Assign responsibility

Cause and Effect Analysis

Introduction

Cause and effect analysis is a systematic way of looking at effects and causes that create or contribute to those effects.

Figure 5: Typical Cause and Effect Diagram

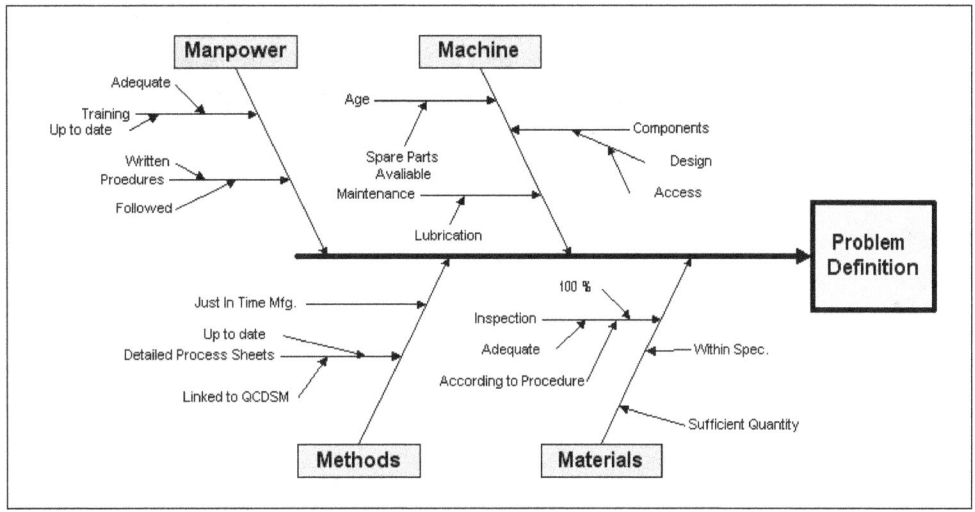

What Are Effects?

Generally, effects are statements about the way things are-descriptions of a problem meeting correction. But effects may also be desired results-that is, how you want things to be when the problem has been solved.

Uses Of The Technique

Thus, cause and effect analysis can be useful at two stages in the problem-solving process:
- Diagnosing the problem
- Implementing a solution

Continued on next page

Cause and Effect Analysis, Continued

Cause and effect analysis is usually shown in a diagram known as a fish bone diagram, because of its shape, or, in less commonly, as an Ishikawa diagram, after its inventor, Dr. Kaora Ishikawa, the Japanese quality-control statistician.

The effect to be analyzed is written to the right on the diagram, in the "fish head". Along each of the "bones", the team records the specific factors that team members considered to be possible causes of the effect. Brainstorming is a common method of identifying these factors.

Figure 6: Example of a Fish Bone Diagram

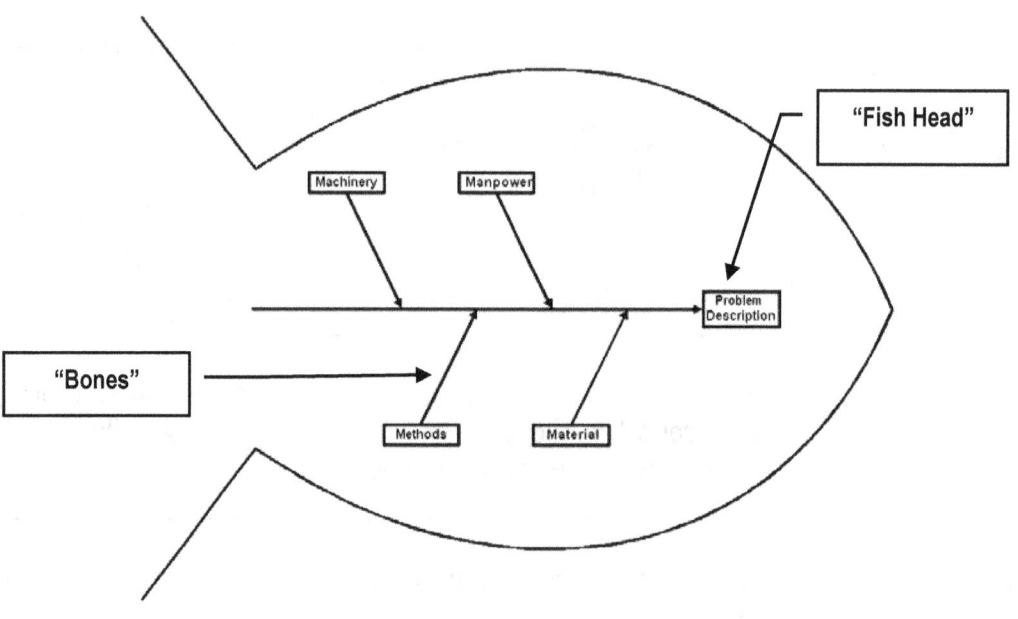

Continued on next page

Cause and Effect Analysis, Continued

Four Categories (4Ms)

To help organize your teams thinking, these factors or causes can be teamed in four categories.

In manufacturing, the four categories are often referred to as the four Ms:
- Machine
- Materials
- Methods
- Manpower.

As an alternative, you can refer to the set of factors as two Ms and two Ps (machine, materials, process, and people.)

Once you've identified all of the possible contributing factors, identifying the most significant factor or factors. To verify that these are indeed the root causes, you should collect additional data to verify that causal relationship.

Diagnosing the Problem

Cause and effect analysis thus helps to diagnose a problem--to trace it back to its root causes. It's the root causes that need to be clearly identified and understood before a team begins to identify and implement solutions.

Identifying the Root Causes

Once the diagram is filled with potential causes, the team scans the diagram for causes that appear repeatedly. Circle that could prove to be root causes.

Continue to search for root causes by asking why? Of each cause contributed in the brainstorming, working back to the causes (or root) of the causes you listed.

Note that the team will still need to verify a likely root causes by collecting additional data on those causes. If the data does not substantiate the likely root causes, the team may have to collect data on some of the less likely causes on the diagram.

Pareto Analysis

Introduction

Pareto analysis is a technique that separates the "vital few" from the trivial many. In solving problems, your team can use Pareto analysis:
- To narrow the list of potential causes
- To determine the priority causes that should be addressed first

Figure 7: Typical Pareto Chart

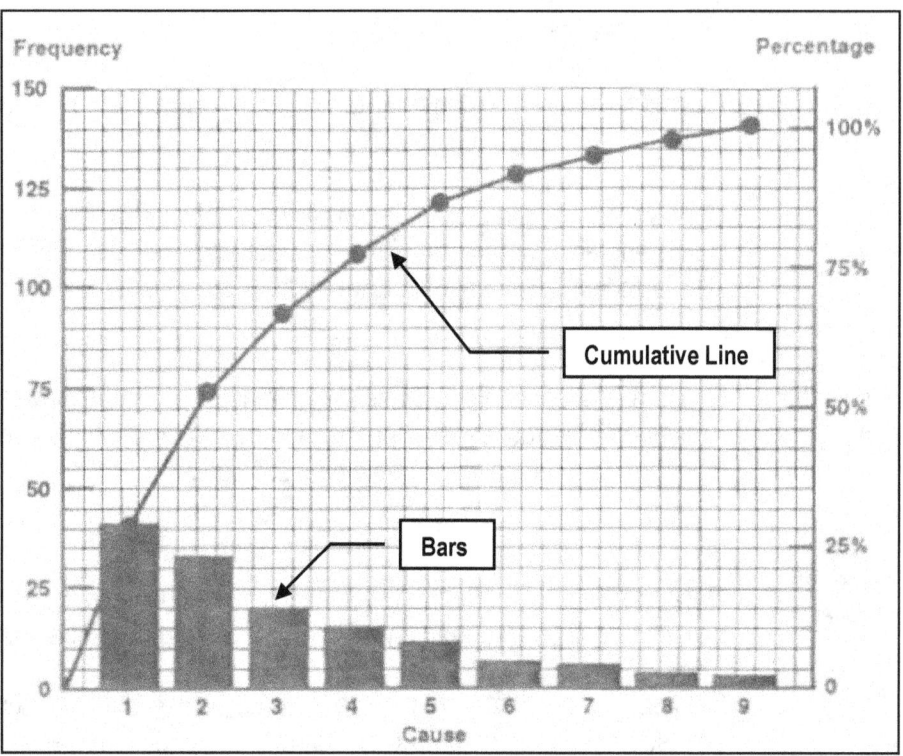

Once you have collected some data about possible problems or causes, a Pareto diagram as shown in Figure 7, lets you examine those data in a systematic way. Pareto analysis ranks data, presenting them in a bar graph. By revealing which problems are the greatest or which causes are the most important, a Pareto analysis can help a team set its priorities for action.

Pareto analysis, named for 19th – century economist Vilfredo Pareto, is designed to point out inequalities. The familiar 80 – 20 rule – that 80 % of our business comes from 20% of our customers – is an example of Pareto analysis.

Continued on next page

Pareto Analysis, Continued

Step 1: Data Gathering and Computation	The first step in analyzing the causes of a problem is to plan the data gathering. • Who will gather the data? • Using what method? • For how long? • How will you quantify the information?
Standard Unit of Measure	To gather meaningful data, examine the possible causes to determine a standard unit of measure that will let you compare causes. For example: in a print shop, a standard unit of measure might be the number of dollars lost in jobs turned away when the copier is unavailable for a particular reason. Other measures include hours of machine downtime, or number of copies postponed. The easiest measure for this example is the number of times a given cause occurs, also called the *frequency*.
Step 2: Tallying and Transferring Data	One of the more common devices for tallying data is the check sheet as shown in Figure 8. Note that tentative causes, that were listed earlier on a flipchart, are stated so that copier users can make discrete choices when reporting. For instance, "machine not operating" has been accounted for with two check sheet questions, number 8 and 9, which will help distinguish between causes that might be affected by training and those that might only be remedied by replacing the copier or establishing a service contract.

Continued on next page

Pareto Analysis, Continued

Figure 8: Typical Check Sheet

Figure 8: Typical Check Sheet

262	1. I had **no problem** or delay getting the copies I wanted
6	2. I made my copies, but they were **not good quality**.
12	3. I couldn't make copies because there was **no copier paper**.
4	4. I couldn't make copies because there was **no toner**.
41	5. There were other users ahead of me, but I had to **wait less than 5 minutes**.
15	6. There were other users ahead of me, and I had to **wait 5 or more** minutes.
7	7. I had a **priority job**, and I had to wait behind other users or break in line to get my job done.
20	8. The copier malfunctioned while I was using it, but **I was able to fix** it using the instruction cards.
33	9. The copier was not operating or malfunctioned while I was using it, and **I was unable to fix** the error.
3	10. **None of the above** describes the problem I had with the copier.

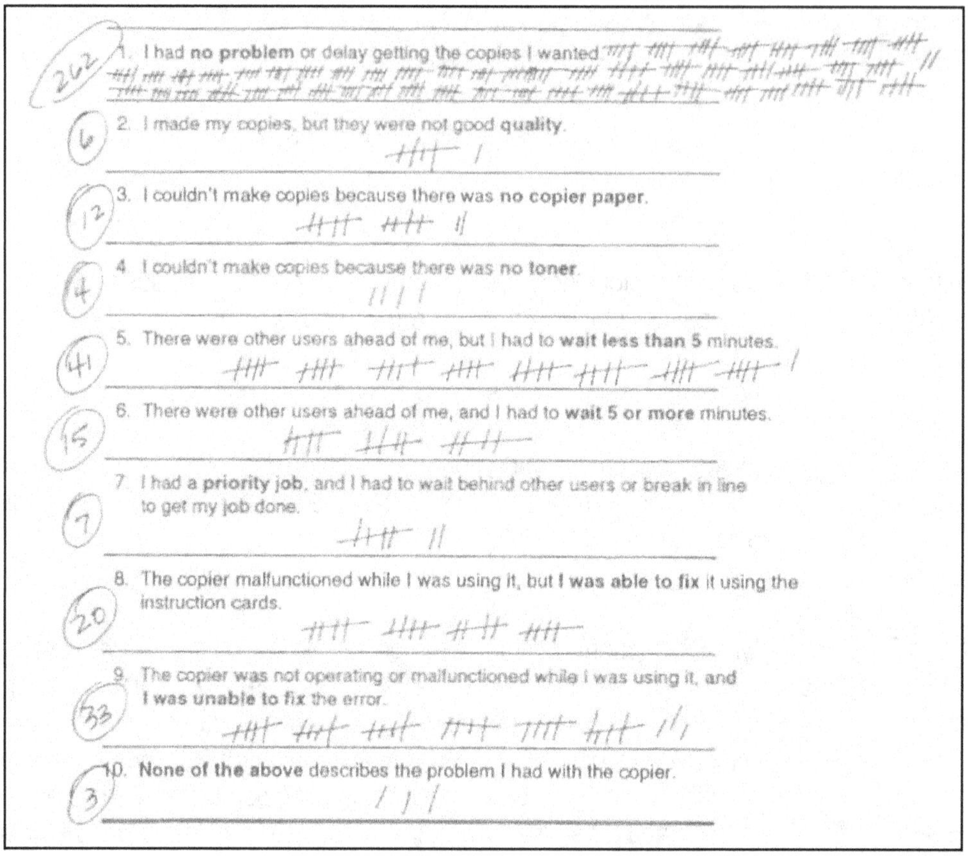

Step 3: Graphing the Data

Now you are ready to display the data in a Pareto diagram. Figure 7 shows a partially completed Pareto diagram

Vertical and horizontal axes have been drawn. The vertical axis is labeled with the chosen standard unit of measurement – frequency. The scale should run from 0 up to the total number of occurrences. (When you are doing a Pareto diagram for a Reliability Team problem, you will find it helpful to work on graph paper.)

Continued on next page

Pareto Analysis, Continued

Plotting the Bars	From left to right, along the horizontal axis, you will construct a series of bars, one for each cause listed in the steps 2 data table, starting with the most frequently occurring cause and ending with the least frequently occurring. Each bar's height corresponds to the frequency of occurrence of the cause it represents.
Drawing the Cumulative Line	The cumulative line is an "overlay" on the bar graph. At the right side of the graph there is a second vertical scale, running from 0 to 100%, with the 100% point at the same height as total occurrences. Using this scale, plot a point above each bar for the cumulative percentages.
Interpreting the Diagram	The bars alone show the relative importance of each cause. The cumulative line makes more graphic the idea of what proportion of a problem's occurrence are attributable to which causes. The first two causes together account for 74 of the observations – 52% of the copier problems observed during the two week period. With such a diagram, your team can see where to focus its energies in developing solutions.

Force Field Analysis

Introduction	Solving problems involves changing the status quo, closing the gap between the problem and the solution, between the present state and some desired future state. Though your team may have been set up to help your organization make a change, there nevertheless may be organizational and individual needs and forces hindering that change.

Force field analysis, a tool developed by organizational researcher, Kurt Lewin, can be used to identify the forces that help and those that hinder a change. |
| **Purpose** | Force field analysis is a method of addressing both the *driving* and the *restraining* forces coming into play when your team attempts to implement a solution to a problem. It thus provides a tool for troubleshooting various solutions and the strategies for putting those solutions in place. |
| **Use Of Diagram** | The analysis is done with a diagram, a graphic techniques that helps:
• Trigger ideas
• To record brainstormed ideas
• To illustrate the strengths of driving and restraining forces
• To reveal any need to amend a proposed solution
• To consider driving and restraining forces in developing implementation strategies. |

Continued on next page

Force Field Analysis, Continued

Figure 9: Typical Force Field Diagram

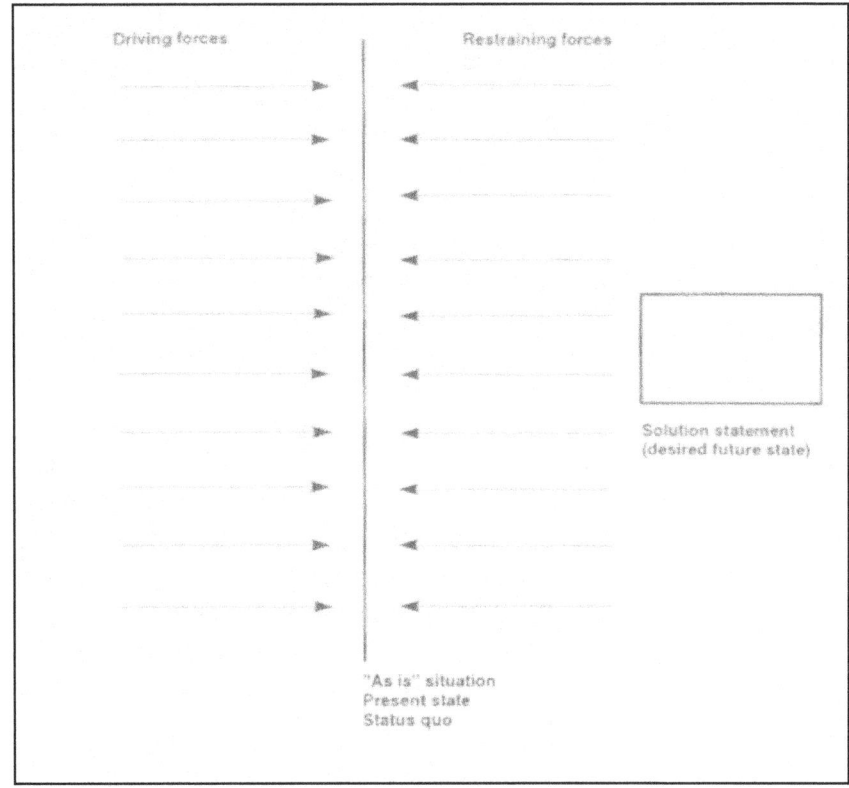

Explanation of Diagram

The vertical line in the center of the force field diagram represents the status quo – what currently exists, the situation as it is.

The box in the lower right of the diagram contains the solution statement, the desired state the team is trying to reach.

The arrows on either side of the vertical line and directed toward it represent opposing forces, which account for the status quo line being where it is.

Continued on next page

Force Field Analysis, Continued

Changing the Equilibrium

The *driving* forces on the left and the *restraining* forces on the right are in balance. But the equilibrium can be shifted in the direction of positive change in one of three ways:
1. By strengthening or adding new driving forces.
2. By reducing or removing some of the restraining forces.
3. By changing the direction of the restraining forces.

Three Key Principles

Note three key principles for changing forces:
1. Concentrating solely on strengthening or adding driving forces often results in *increasing* the tension, which in itself may reduce the anticipated benefit of the proposed change.
2. Attempting to induce change by removing or eliminating restraining forces will generally *reduce* tension and effect more stable changes. Restraining forces that are removed cannot push for a return to old ways of doing things.
3. One of the most efficient ways to make change happen is to change the direction of one of the forces. Turning around a force perceived as restraining provides great impetus toward the desired state. For example, management disapproval may be one of the restraining forces a team perceives it has to counter. Engaging management commitment and support turns that restraining force into a driving force.

Defining the Current Situation

Overview

Introduction

When addressing chronic loss, the Reliability Team must have an awareness that action is required, This typically comes about when there is noticeable deviation between what we expect to happen (Should) and what is in fact happening (Actual) as shown in Figure 10.

Figure 10: Recognition of a Problem

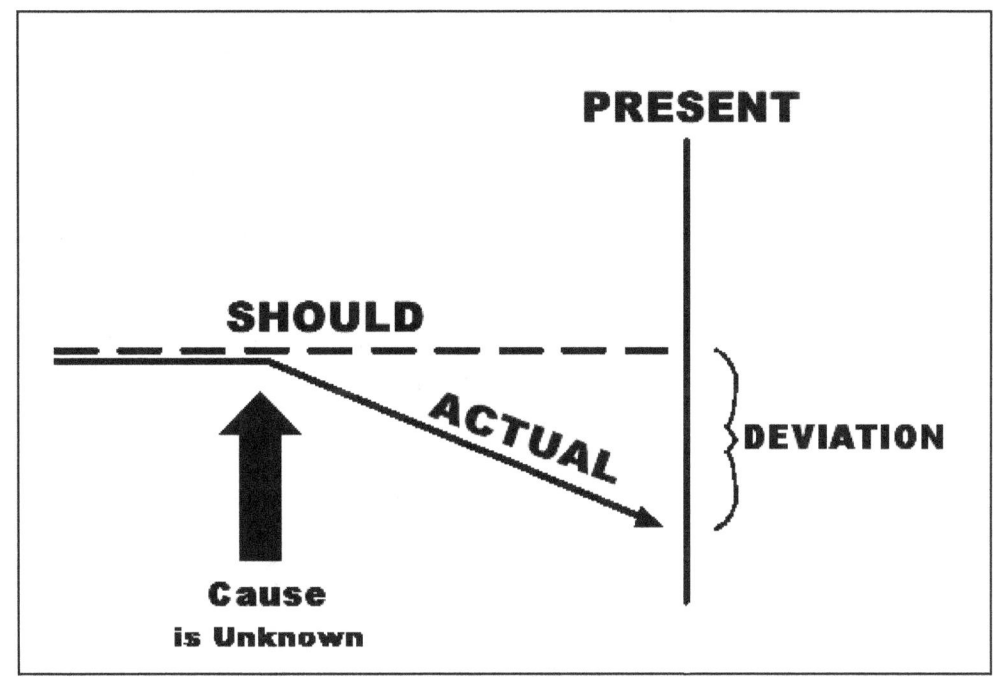

In This Chapter

The topics covered in this chapter include:

Topic	See Page
Workplace scan	46
Identifying basic operating principles	56
Identifying operating standards	58
Identifying Interacting elements	59
Quantifying physical changes	60

WORKPLACE ANALYSIS

Introduction

A workplace analysis gives you a clear picture of the current situation in the target area. It is designed to show the physical structure and design of the target area through the use of an area map. The workplace analysis should show the flow of people, equipment, and materials through the target area

Purpose

The Workplace Analysis is the foundation piece of Focused Operations Performance Improvement efforts. The purpose is to help the Reliability Team clearly understand the current conditions which exist in the target area. The workplace analysis is designed to provide facts and clarify any prior assumptions that the team may have about the target area.

Description

The Workplace analysis includes five basic elements:
- Definition of the target area and a list of purpose and functions
- An area map and arrow diagram
- Workplace diagnostic check list
- Photographs of the target area posted on a problem chart
- A Workplace analysis display

In this Section

The topics described in this section are located as indicated below

Establish the Current State Map

Introduction   This unit shows the participant how to establish a Current State Map

Description   Mapping begins at the "Door-to-Door" level in your plant where you draw process categories like: mixing, machining, baking, packaging instead of recording each process step. You can change the level of magnification and zoom in to map every individual step in each category, or zoom out to encompass the value stream external to your plant.

Figure 11:
Typical Current
State Map

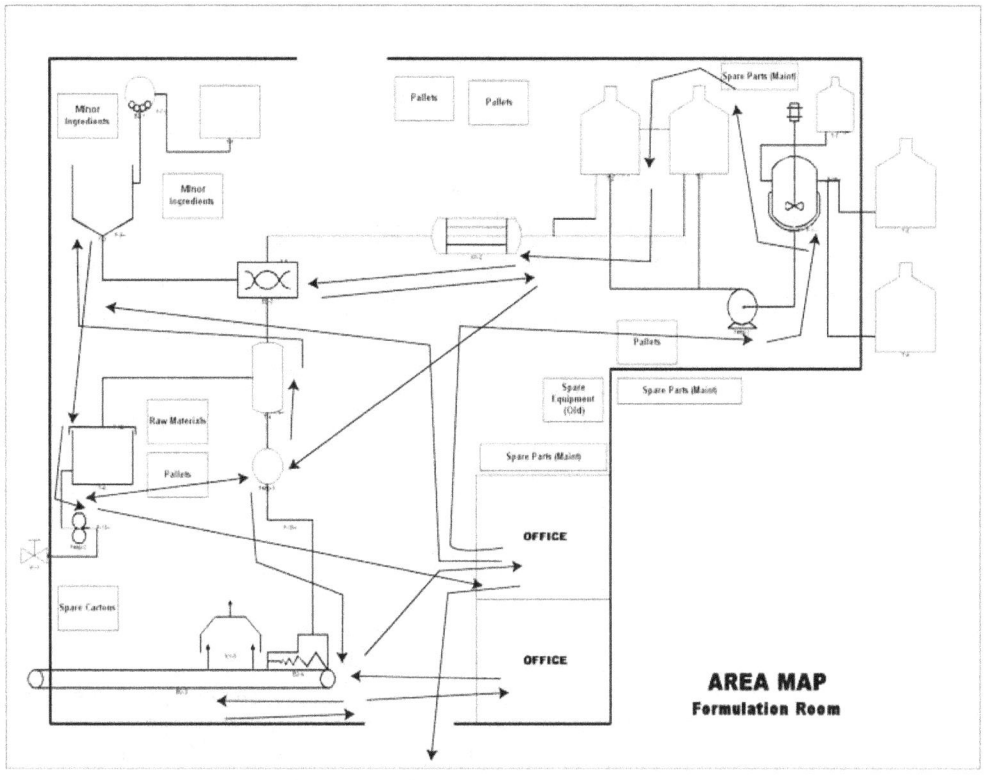

Purpose   The purpose of the current state map is to show you the overall flow through the plant. The point is not the map, but to understand the flow of information and material.

Continued on next page

Establish the Current State Map, Continued

Procedure	Use the following procedure develop a current state map:

Step	Action
	Developing Flow Diagram
1	Collect current – state information while walking along the actual pathways of material and information flows yourself.
2	Start at the end of the process (not the beginning) and work up stream. This way you will begin with the processes linked most directly to the customer. Use a stop watch to determine actual times – do not rely on standard time or information that you do not personally obtain.
3	Use a Yellow Post-It Note to identify each part of the process. Write the name and/or draw the symbol on the post-it note and attach it to the paper on the wall.
4	Continue to the previous step in the flow (remember, you are working backwards.
5	Do these steps until you have the entire flow diagrammed.
	Data Gathering
6	Go out to the target area and look for amounts of inventory, waste, unsafe conditions, and other problems.
7	Observe each step of the process to determine the amount of product flow and time to produce it.
8	Indicate the amount of inventory on a yellow Post-It.
9	Write in the amounts and time on a yellow Post-It note with a data box drawn on it. Place it below the appropriate symbol in your flow diagram.
10	Using a post-it of a different color, identify examples of waste and place the post-it beneath the appropriate symbol.
11	Check you current-state map against the actual area to ensure its accuracy.

Continued on next page

Establish the Current State Map, Continued

Procedure **(continued)**

Step	Action
	Gather Information Flow Data
12	Go back to the target area and determine how information flows to each step of your flow diagram.
13	Using a different color Post-It note, diagram the flow of information from the end of the current-state map.
14	Verify the correctness of the information.
15	Your current-state map is now complete.

Create the Opportunity Chart

Introduction

There are four main features to the opportunity chart method:

- Taking photographs of workplace problem areas
- Returning at set intervals to take a new round of pictures of the same areas
- Focusing the camera on the same targets in each round of photographs

Shooting each round of photographs with the camera in the same position (same height, angle, and distance from each target).

**Figure 12:
Typical
Opportunity
Chart**

Continued on next page

Create the Opportunity Chart, Continued

Purpose   The Opportunity Chart is a means of capturing opportunities for improvement in the workplace using photographs to inform and inspire team-based workplace improvement activities.

Procedure   Use the following procedure to develop an Opportunity Chart

Step	Action
1	Inspect the Current State Map and make a list of potential target areas, such as messy or potentially dangerous areas that could be improved.
2	Working with this list, choose subjects for the Opportunity Chart.
3	One team member takes a picture of each subject.
4	.As each area is photographed, team members mark the position of the photographer's feet on the floor and note the camera position and focus relative to the subject.
5	Make a numbered list is of each picture using an OP log sheet
6	Organize these initial photos and attach them to a workplace Opportunity Chart.
7	Print three copies of each picture and used as follows: One copy is stored in an archive A second copy is attached directly to the Workplace Opportunity Chart The third copy is used on the VFP Chart.
8	Use transparent tape or push pins to attach the Workplace Opportunity Chart to a stiff display board. .
9	Place the Workplace Opportunity Chart in a designated area where the team meets

Prepare a Visual Feedback Chart

Introduction

This unit explains the process for creating a Visual Feedback (VFP) Chart as shown in Figure 13.

**Figure 13:
Typical Visual
Feedback Chart**

Purpose

The VFP chart has two goals:
    To make photographic records of current conditions
    To take s time lapse series of photographs of the same target location

Continued on next page

Prepare a Visual Feedback Chart, Continued

Record Conditions that are Hard to Quantify

Disorderly mounds of materials and dirt are hard to quantify. Different people have different rating scales and what may look good to one person may look marginal or unsatisfactory to another. Photos capture the conditions as they were at the time of the photo, allowing people to make their own assessment at a later date.

Recognition of Subtle Changes

Often improvements consist of a series of small changes that are important but hard to see. A series of photographs mounted on a VFP chart helps pinpoint subtle changes that might otherwise be difficult to recognize.

Procedure

Use the following procedure to create and maintain a VFP Chart

Step	Action
	Initial Preparation
1	Prepare supplies
2	Select photos from the Opportunity chart as indicated in step 7 of the Opportunity Chart procedure.
3	Glue one set of each photo in the round 1 column of your chart.
4	Label the picture and supply information on location, date, target owner, etc.
5	Display the chart
	Adding Photos
6	Take a second round of photos, using the VFP log to make sure that pictures are taken from the same location and point of view. Take several shots of each target.
7	Review the pictures as a team and select best pictures to represent each target.
8	Attach the new picture in the second round column.
9	Re-display the VFP Chart
10	Follow steps 6 – 10 for rounds 3 and 4. If a target improvement is completed before round 4 take a final picture of the improved target and make a notation in the next available column, in the spot where the picture would normally go.

Create the Future State Map

Introduction	This unit shows the participant how to create a Future State Map. Figure 14 shows the target area after items have been removed or organized.

**Figure 14:
Typical Future
State Map**

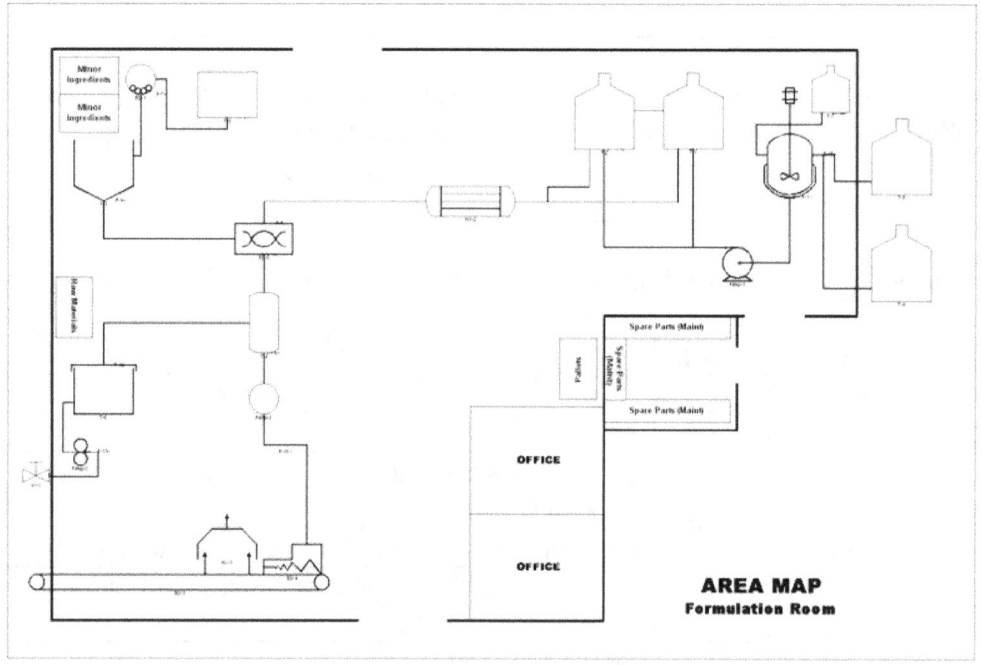

Description	The future state map is similar to the current state map; it also begins at the end of the process and works backwards. The difference between the two is that you have removed most or all of the waste.

Purpose	The purpose of a future state map is to build a chain of production where individual processes are linked to their customer(s) either by a continuous flow or pull, and each process gets a close as possible to producing only what its customer(s) need when they need it.

Continued on next page

Create the Future State Map, Continued

Procedure   Use the following procedure to create a Future State Map

Step	Action
Demand	
1	Determine Takt Time for downstream processes.
2	Determine if you need to build a finished goods supermarket from which the customer pulls, or direct to shipping.
Material Flow	
3	Determine where you can use continuous flow processing.
4	Identify where you will need to use supermarket pull systems to control production of upstream processes.
Information Flow	
5	Determine at what point in the production chain you will schedule production. Keep in mind that all material transfers downstream of the pacemaker process need to occur as a flow.
6	Determine how to level production mix at the pacemaker process.
7	Determine the increment of work you will consistently release and takeaway at the pacemaker process.
Supporting Improvements	
8	Determine the process improvements that will be necessary for the value stream to flow as the future state design specifies.
9	Starting at the end of your new process, diagram the process flow.

CONDUCTING A PHYSICAL ANALYSIS

Overview

Introduction

Reliability Teams should only conduct one Physical Analysis for any given phenomenon. If the team comes up with more than one Physical Analysis, they have generally failed to properly define the phenomenon, typically in this case, it will be the result of too broad a definition.

Understanding the Changes

Physical analysis examines each of the machine and process inputs along with other physical elements and the phenomenon in order to understand the change that has occurred. All that remains is the identification of the interacting elements and quantifying the physical changes that occur as a part of the phenomenon.

Description

A Physical Analysis consists of four steps:
- Identify basic operating principles
- Identify operating standards
- Identify interacting elements
- Quantify physical changes

In This Section

The topics covered in this section include:

Identifying Basic Operating Principles

Introduction	All equipment runs according to a set of basic operating principles. This occurs regardless of whether it is running in manual, semi-automatic, or automatic mode.
Importance	Most equipment operators understand how the main unit operates, but generally know very little about peripheral mechanisms such as pneumatic systems, hydraulic systems, sensors, and electrical systems. Very few understand the control or measurement principles. If operators are on the team (and they should be), they must have an understanding of the operating principles or they will not be able to conduct a proper Physical Analysis
Sources of Information	The operations manuals, maintenance manuals, and drawings are a good source of this information. Vendor representatives are another good source of this information. All Reliability Team Members should refer to these sources before conducting a Physical Analysis.

Identifying Operating Standards

Introduction	The next step is to identify the operating standards associated with normal (defect-free) operating principles. It is critical to ensure that the proper standards are identified for each mechanism.
Standards Identification Process	There are two steps to identifying operating standards: • Relate the operating principles to equipment mechanisms • Identify how those mechanisms normally function in order to prevent the phenomenon from occurring.
Example of Operating Principles	In the distillation process, if the temperature of the crude is allowed to rise above 900°F, a sudden change in the shape of the distillation curve takes place. As the temperature goes from 900° to 1100°F, the cumulative volume recovered would exceed 100%, and crude would still be boiling in the vessel. As the complicated hydrocarbons remaining in the vessel that have not vaporized at 900°F are heated to higher temperatures, the energy transfer from the heat is enough to crack the molecules into two or more smaller molecules.
Example of Related Standards	Temperatures in excess of 900°F are to be avoided in the distilling process. This may be accomplished through vacuum flashing in order to reduce the pressure and thereby reduce the point at which vapor will be formed. *Standard: Distillation processes will operate at temperatures less than 900°F* Use of the CAT Cracking process allows for heat at or above 900°F. The CAT (Catalyst) Cracking process is designed to break up the large molecules into smaller useable molecules while minimizing the resulting coke production *Standard: The CAT Cracking process will be used to process the heavy cuts of crude at temperature near or above 900°F*

Focused Operations Performance Improvement

Identifying Interacting Elements

Introduction

Equipment problems may result from a combination of malfunctions depending on the complexity of the equipment. It is important therefore, to identify the interacting elements that are involved and what changes might occur as a result. Interacting elements include:
- Equipment
- Tooling
- Work in process (product)

Identifying Interacting Elements

An example of interacting elements is shown in the center column of Table 2. The Reliability Team should make a sketch of how these interacting elements move in relation to each other. The sketch becomes a "Contact Diagram" and will be valuable in understanding the mechanics behind the phenomenon and in creating a Contact Sheet.

Table 2: Relationship of Interacting Elements and Phenomenon

The following contact sheet for a roll grinder in a paper mill demonstrates the relationship between phenomenon and interacting elements.

Abnormal Phenomenon (Contact Diagram)	Interacting Elements	Quantitative Physical Change
ROLL GRINDER	Between Roll center of rotation (A) and Axis of moving cutter (B)	• Distance (c) • Is not parallel • Does not produce specified cylindrical shape
ROLL GRINDER	Between the Roll Center of rotation (A) and edge of cutter (B)	• Distance (c) • Fluctuates, so • Finish dimensions of roll are not constant.

Quantifying Physical Changes

Introduction	Finally, the Reliability Team has to consider what measurable changes occur when the phenomenon occurs. The team has to determine the appropriate measurement units needed to quantify the change in each pair of correlating elements related to the phenomenon
Process of Quantifying Physical Changes	The Reliability Team must now review the physical quantity or quantities linking the elements. For example, if your team was looking at the Roll Grinding process in a paper mill, it would quantify the relative positions of the grinding wheel and the paper machine roller being resurfaced. The quantification would be the distance between the roll and the grinding wheel at key spots along the length of the roll.
Clarifying the Changes	The team would clarify how the changes occur by summarizing them in a written explanation or in a final diagram. The written explanation is probably the most useful to people outside of the reliability team. In some cases you may find it hard to express what is occurring in words, but a diagram or series of diagrams my make that task easier.

Problem Analysis

Overview

Introduction

Problem analysis or cause-and-effect analysis (also known as Factor Analysis) is the process of identifying which factors cause a given problem. Reliability Teams use cause-and-effect fishbone diagrams to conduct factor analysis. This helps to determine which factors contribute the most to producing a particular defect and establishing it as a target for improvement.

Figure 15:
Typical Cause &
Effect Diagram

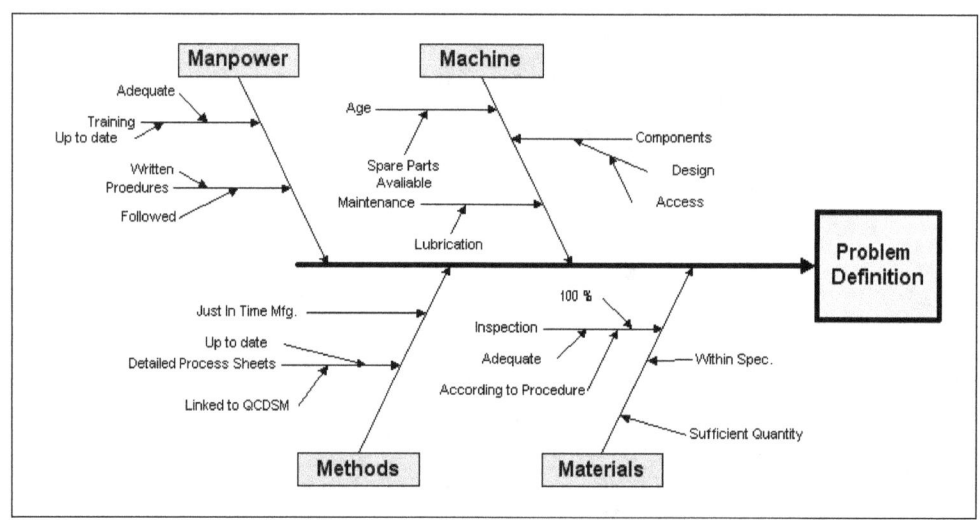

In This Chapter

The topics covered in this chapter include:

Topic	See Page
Relationship of Factors and Causes	63
Limits of Traditional Problem Analysis Techniques	64
Analysis Tools	67

Relationship of Factors and Causes

Introduction

It is important to distinguish between a factor and a cause.

Definitions

The following definitions will be used throughout this chapter and the entire book:

Term	Definition
Factor	Any condition that may potentially contribute to a given phenomenon.
Cause	Any factor that immediately precedes the phenomenon and always produced it. A factor can only be a cause if it is capable of producing the phenomenon by it self.
Causal Factor	Any factor that contributes to the phenomenon but may or may not directly produce it.

Example

If a dough machine is experiencing a high rate of scrap at start up, a Reliability Team might study the problem using a cause and effect diagram. The items that the team assigns to the Manpower vane might include:

- Operator error
- Lack of experience or skills
- Quality consciousness
- Insufficient knowledge
- Fatigue
- Insufficient training

These would all be examples of factors – any one or several could actually be the cause of the problem. At the beginning you may not have enough information to determine a cause yet.

If upon further study, you determine that the operator was not properly trained you would narrow the cause down to "Operator Error due to insufficient training". The cause then would be insufficient training.

Limits of Traditional Problem Analysis Techniques

Introduction   The traditional cause-and-effect diagram is an easy to sue and somewhat effective tool for use at the shop floor level, but it has limitations when it comes to addressing chronic losses that block the path to true performance improvement.

Approach   In factor analysis the Reliability Team is encouraged to think up potential causes without studying mechanisms, structure or component configurations and functions. This may lead to the omission of some casual factors which may result in weak linkage between the phenomena and a given factor.

Flaws in the Problem Solving Process   The following are typical flaws in the problem solving process:
- **Crude methods for identifying factors:** i.e. Timing (of what?) Pushrods (which ones?)
- **Inclusion of unrelated factors:** i.e. Routine and daily checks, parts replacement, materials, deformation (How are these related to the phenomenon?)
- **Omission of causal factors:** i.e. Center line of pushrods (friction damage) vertical and horizontal alignment of cam, loose roller guides, etc.
- **Logical inconsistency between factors:** i.e. abrasion, wobble (how can something be too tight and too loose at the same time?)

Reasons For Flawed Analysis   There are a number of reasons for flawed analysis:
- The machine is poorly understood
- Phenomenon are not considered carefully and logically
- Operating principles are not understood
- Factors are not understood in sufficient detail
- Emphasis on prioritization ignores all but the most influential factors.

Continued on next page

Limits of Traditional Problem Analysis Techniques, Continued

Figure 16: Cause & Effect Diagram Example

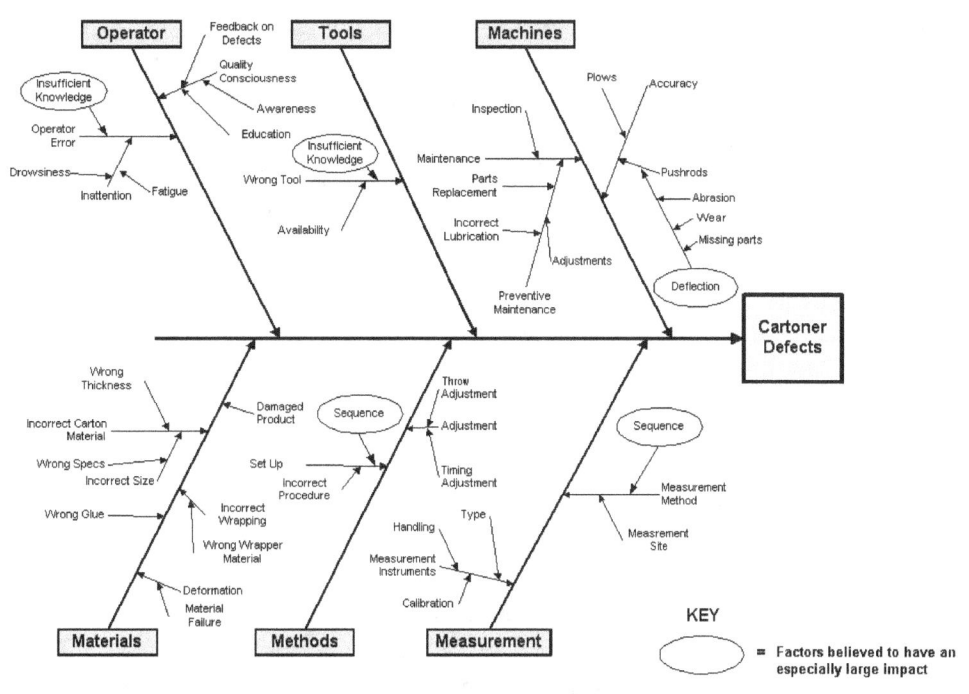

Machine is Not Properly Understood.

Figure 16 show an example doesn't show what machine operations were addressed. A guess must be made as to what was happening in the cartoner. Insufficient knowledge, sequence and deflection are indicated as possible factors; however other things could be occurring inside of the machine that is not operator related. In this case the analysis would most likely be incorrect and miss the mark. This is very common with the conventional problem solving approach.

Phenomenon Not Considered Carefully and Logically

People employing traditional problem solving analysis do not spend much time looking at phenomena in a careful organized manner. They typically oversimplify the phenomenon in general terms and then begin brainstorming factors. This allows unrelated factors to be considered while causal factors are overlooked. This leads to too narrow a scope of factors which leads to improper conclusions based on insufficient information.

Continued on next page

Limits of Traditional Problem Analysis Techniques, Continued

Operating Principles Not Understood

The design of all machinery is based on specific operating principles. Operating principles are physical explanations of how something works. Operating principles are the foundation upon which a machine's mechanisms and structure are designed. Too often, people attempt to analyze machine problems without attempting to understand the principles of how that piece of equipment operates and how the components and mechanisms interact. They fail to understand the relationship between controls, measurement systems, and mechanisms. When this occurs, important factors are typically left out of the analysis as shown in Figure 16.

Factors Not Understood In Sufficient Detail

Other problems with the analysis occur when factors are selected. It is common for Reliability Teams to lump items under overly-broad categories, this leads to considering only major units or mechanisms. This in turn inhibits detailed analysis. In some cases, some mechanisms may be broken down to the component level while others are completely ignored.

When Reliability Teams consider factors only at the equipment of mechanism level, they end up missing a lot data because they are too general. The team may find that it needs to break down the major categories even further into component characteristics such as:
- Strength
- Shape, and texture
- Dimension accuracy
- Means of attachment

Prioritization Ignores Most Influential Factors

We are taught to prioritize things for an early age; it is not surprising that we carry this through to our careers as adults. Typically a Reliability Team begins their task by identifying what they consider the most influential factors and focus their efforts accordingly. The distinctions they make are quite often causing them to miss important factors.

Need for Better Approach

Despite the weaknesses typically found in traditional problem analysis techniques, they are often very useful and effective in limited circumstances. The main weakness is that while helping to significantly reduce defects, it cannot eliminate them. A better approach is needed.

P–M Analysis Approach

Introduction

P-M Analysis is more than an improvement methodology. It is a different way of thinking about problems and the context in which they occur. P-M analysis helps eliminate chronic losses in three stages:

- Look at phenomena analytically and systematically
- Review all causal factors
- Identify all abnormalities and reduce them to zero

The "P" and "M" in P-M analysis do not stand for preventive or productive maintenance. The "P" stands for "phenomenon"—the abnormal event to be controlled. It also stands for "physical"— the perspective we take in viewing the phenomenon. "M" refers to "mechanism," and also to the four production inputs (4Ms) we examine for causal factors: machine (equipment), man (people), material, and method. The term "mechanism" applies to any grouping of equipment elements (including jigs and tools) with a common function. It also refers to the "mechanics" or failure mode of an abnormal event.

Purpose

P-M analysis *physically analyzes chronic losses according to the inherent principles and natural laws that govern them.* This analysis clarifies the mechanics of their occurrence and the conditions that must be controlled to prevent them.

Basic Principle

The basic principle behind P-M analysis is to first understand—in precise physical terms—what happens when a machine breaks down or produces bad parts or material, and how it happens. Only then can we identify and address all causal factors and thus eliminate the chronic loss.

Continued on next page

P–M Analysis Approach, Continued

P–M Analysis Sequence

Looking at it another way, P-M analysis is a refined variation of cause-and-effect analysis that considers all causal-factors instead of trying to decide which are most influential. Teams using P-M analysis follow this sequence:

Step	Action
1	Physically analyze chronic problems such as defects and failures according to the machine's operating principles
2	Define the essential or constituent conditions underlying the abnormal phenomena.
3	Identify all factors that logically contribute to the phenomena in terms of the 4Ms: • Equipment mechanisms • Materials • Methods used • People's actions.

Continued on next page

P–M Analysis Approach, Continued

Figure 17: Operations Analysis Flow Chart

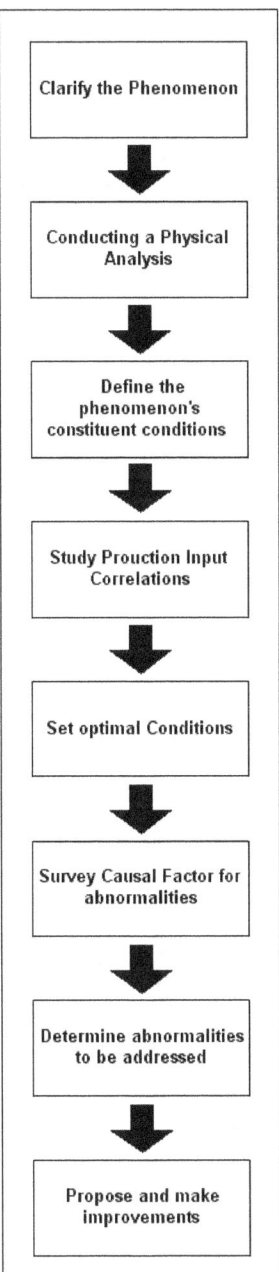

Clarify the Phenomenon	Carefully define and categorize the abnormal occurrence.
Conducting a Physical Analysis	Describe the phenomenon in physical terms. e.g., How the process conditions change over time to produce the error or defect
Define the phenomenon's constituent conditions	Identify all of the conditions that will consistently produce the phenomenon
Study Prouction Input Correlations	Look for potential cause and effect relations between conditions and machine, materials, methods, and manpower.
Set optimal Conditions	Review the equipment's current precision levels to determine where standards are deficient.
Survey Causal Factor for abnormalities	Confirm which factors identified in step 3 and 4 exhibit deviating conditions.
Determine abnormalities to be addressed	Review survey results and list all abnormalities to be addressed. (Include minor defects)
Propose and make improvements	Implement a corrective measure or improvement for each abnormality, and then create appropriate operating standards and Preventive Maintenance procedures to maintain optimal conditions.

Continued on next page

P–M Analysis Approach, Continued

Clarify the Phenomenon

A phenomenon is a fact or occurrence that can be observed. When conducting P-M Analysis, the Reliability Team uses their collective observation skills to consider and comprehend facts and concepts. This step is critical to ensure that critical information is not omitted.

Observing the Phenomenon

Unfounded assumptions can create an obstacle to obtaining the proper solution. The Reliability Team members must verify the phenomenon with their own eyes. Merely "looking" however is not sufficient, the team should ask the following questions when observing phenomenon:

- When specifically does the phenomenon occur? (What operation?)
- Does it consistently occur under the same circumstances?
- At what intervals does it occur? (regular or irregular)
- What is the impact of time on the phenomenon? (does it get better or worse)
- Where does it occur? (more than one machine)
- Does the operator affect the phenomenon? (which one(s))

Level of Specificity is Important

The team must avoid vague descriptions such as "many operating defects occur". When a phenomenon occurs, the team must define it in very specific terms such as "The bagger is cutting every fifth bag short and damaging the product as well as creating an ineffective seal." Or "the Case Packer places an extra row of product in the first 20 cases produced after changeover."

Defining Phenomenon

It is important to understand and sort phenomenon based on type or pattern observed, how they occur and where they occur. The critical steps in this process include:

- Eliminate pre-conceived ideas or notions
- Carefully observe and analyze the facts at the point of occurrence
- Sort and stratify the phenomenon thoroughly
- Compare normal with abnormal out put to pinpoint significant differences

Continued on next page

P–M Analysis Approach, Continued

Classifying Phenomenon

Use the following questions to stratify the results of observations and classify phenomenon:

Type	Question
Who	Is there any variation among people involved in the operation? (shift, new operators, temp employees, etc.)
What	Is there any variation caused by production materials? (material differences, part dimensions, shapes, structure)
Where	Is there any variation due to equipment or components? (process, machine types, jigs, fixtures, etc)
When	Is there any variation related to time factors? (shift, hour of day, day of week, etc.)
Which	Is there any variation trend over time? (increase or decrease of problems, changes in nature of problem, seriousness, etc.)
How	Is there any variation in the circumstances leading to the occurrence? (occur frequently or rarely, abrupt or gradual onset, continuous or intermittent, regular or irregular intervals, etc.)

Conduct a Physical Analysis

A physical analysis explains properly classified phenomenon from a physical stand point. Without a physical analysis, Reliability Teams have a tendency of use experience, gut feeling or intuition, or initial impressions to describe phenomenon. This leads to incorrect solutions being applied to correct chronic problems which may result in making the problem worse or at the very least not providing a long term solution.

Continued on next page

P–M Analysis Approach, Continued

Physical Analysis Procedure

The best approach is to think visually, using drawings and diagrams of how equipment components interact to cause the defect or failure. Use the following procedure when conducting a physical analysis:

Step	Action
1	**Identify the operating principles** – review machine diagrams and manuals to understand the equipment's basic operating principles.
2	**Identify the operating standards** – learn the functions and mechanisms by making simple diagrams or sketches.
3	**Identify the interacting elements** – Draw contact diagrams and ladder logic to identify what relationships define the phenomenon.
4	**Quantify the physical changes involved** – identify appropriate physical quantities and changes in those quantities.

Continued on next page

P–M Analysis Approach, Continued

Table 3: Examples of Steps 2 & 3

The following are examples of operating principles and standards from a physical analysis:

Operation	Operating Principle	Operating Standard
	A bag is created when film is pulled over the filler tube and a fin is created by sealing the two sides of the film by heat. The length of the bag is determined by a two part horizontal heating element and cutting blade mounted between the two elements. The top element forms the bottom of the bag being filled. The bottom element forms the top of the bag previously filled. The blasé separates the two bags.	1. Pull the film over forming shoulders 2. Insert the sides into the vertical forming slot. 3. Adjust the film drive so that it makes firm contact with the film. 4. Advance the film to the end of the filler tube. 5. Close the vertical sealer. 6. Inspect the resulting seal 7. Engage the horizontal sealer. 8. Test run and inspect the sealed bag 9. Turn on filler.
	Cases of product arranged on a pallet are mechanically wrapped using a stretch wrapping film. The pallet is rotated on a turntable while film is wound around the load. A series of rollers applies a predetermined amount of tension to stretch the film which will relax to its original state when tension is released.	1. Place the pallet on the turntable. 2. Stick film between stacks of product 3. Rotate turntable 2 revolutions. 4. Check wrap 5. Press the "Start" button and wrap the entire load. (Film is automatically cut) 6. Remove the load from the turntable.

Continued on next page

P–M Analysis Approach, Continued

Identify the interacting elements

The next step is to look closely at the elements or equipment components that interact to produce abnormal phenomenon. Basically a phenomenon is a cause and effect relationship
 between equipment and products. The Reliability Team should look at "Cause" as the condition of the equipment and at the "Effect" as the quality of the product. If this relationship deteriorates, it will be reflected in product quality.

Table 4: Interacting Elements

Table 4 shows the relationship between abnormal phenomenon and interacting elements.

Abnormal Phenomenon	Contact Diagram	Interacting Elements
Misalignment of shaft		Reduction gear and shaft from 300 hp Electric Drive motor
Improper Length of Bag		Registration mark reader, Seal Clamping Speed Timing (Cam Adjustment)

Quantify the Physical Changes Involved

When the interacting elements have been identified and diagrammed, the team must quantify the physical changes that occur in their relationship. Table 4 shows the link between the phenomenon and quantifiable physical changes.

Need To Understand Mechanisms & Structure

Understanding the mechanisms and structure of the equipment is critical to the Reliability Team. It helps to clarify what is happening to produce certain malfunctions and defects. Conducting P-M Analysis without understanding the equipment will cause the team to overlook key aspects of the problem.

Continued on next page

P–M Analysis Approach, Continued

The Value of Diagrams

Most Reliability Teams will find it difficult at best, to explain the mechanisms and structures of machines and equipment despite the fact that they may use them everyday. Most teams will find it very useful to prepare machine diagrams and drawings when they start a P–M Analysis. This forces the Reliability Team to carefully examine each part of the machine, which typically leads to the discovery of a greater number of hidden abnormalities.

Table 5: Example of Contact Sheet for Roll Grinding Machine

Table 5 is an example of what a contact sheet for a roll grinder in a paper mill might look like.

Abnormal Phenomenon (Contact Diagram)	Interacting Elements	Quantitative Physical Change
ROLL GRINDER	Between Roll center of rotation (A) and Axis of moving cutter (B)	• Distance (c) • Is not parallel • Does not produce specified cylindrical shape
ROLL GRINDER	Between the Roll Center of rotation (A) and edge of cutter (B)	• Distance (c) • Fluctuates, so • Finish dimensions of roll are not constant.

Continued on next page

P–M Analysis Approach, Continued

How to Acquire Machine Knowledge

When attempting to acquire machine knowledge, remember to do the following:
- Read and reread the machine manual until the machine is thoroughly understood
- Prepare your own diagrams at the machine
- Study the process cycle, wiring, hydraulic, and other system diagrams as well.
- Investigate set up and running conditions
- Restore all items out of specification before starting the P–M Analysis.

Define Constituent Conditions

The next step in conducting a P–M Analysis is to review all of the conditions that consistently contribute to the problem. These conditions are known as the constituent conditions that are necessary for the phenomenon to occur. The Reliability Team must consider everything that could possibly cause the phenomenon to occur. These conditions encompass all of the causal factors.

Table 6: Constituent Conditions

Table 6 shows to relationship between constituent conditions in the forum categories

4M Category	Constituent Conditions
Machine (Equipment) precision and reliability	Whenever any part of a machine malfunctions, check for links with abnormal phenomena and for the conditions that give rise to those phenomena.
Methods and Standards	Check for links with physical defect phenomena whenever designated standards are in adequate or too lax
Manpower (People)	Check for links with abnormal phenomenon when people charged with adhering to standards do not do so.
Material quality from previous processes	Check for links with abnormal phenomena when materials or parts from previous processes are of poor quality.

Continued on next page

P–M Analysis Approach, Continued

Procedure for Checking Constituent Conditions

The following procedure is helpful when checking each of the 4 Ms to see whether off-standard conditions may be linked to defect phenomena:

Step	Action
1	Identify the functional units which constitute the equipment.
2	Examined each mechanism in and identify the role it plays in relationship to the equipment as a whole.
3	Checked for links with abnormal phenomena when any mechanism, subassembly, or component fails to play its expected role.
4	Investigate links between the appearance of abnormal phenomena in the condition of mechanisms potentially connected to such phenomena.
5	To say if abnormal phenomena appear (even if equipment mechanisms are fulfilling their expected roles) when: • Designated standards are too lax or inadequate • People charged with adhering to those standards are not doing so.
6	Even when conditions involving the equipment, standards, and people's adherence are all in order, check for physical defect phenomena resulting from quality defects in incoming material.

Deriving Causal Factors from 4Ms

The key to successful PM–Analysis is identifying cause and effect relationships between constituent conditions and specific 4M elements. This means that you will need to identify all possible elements to generate the constituent conditions.

Continued on next page

P–M Analysis Approach, Continued

Levels of Causal Factors

There are 3 distinct levels of causal factors:
- Constituent Condition – Unit level
- Primary 4M Correlation – Subassembly level
- Secondary 4M Correlation – Component level
-
- As shown in Figure 18, the constituent conditions are identified at the mechanism (unit) level. Primary and secondary correlations stepped down to the subassembly and component levels respectively. Keep in mind the number of levels can very; for instance you may have five or six levels in the case of more complex machines.

Figure 18: Three Levels of Causal Factors

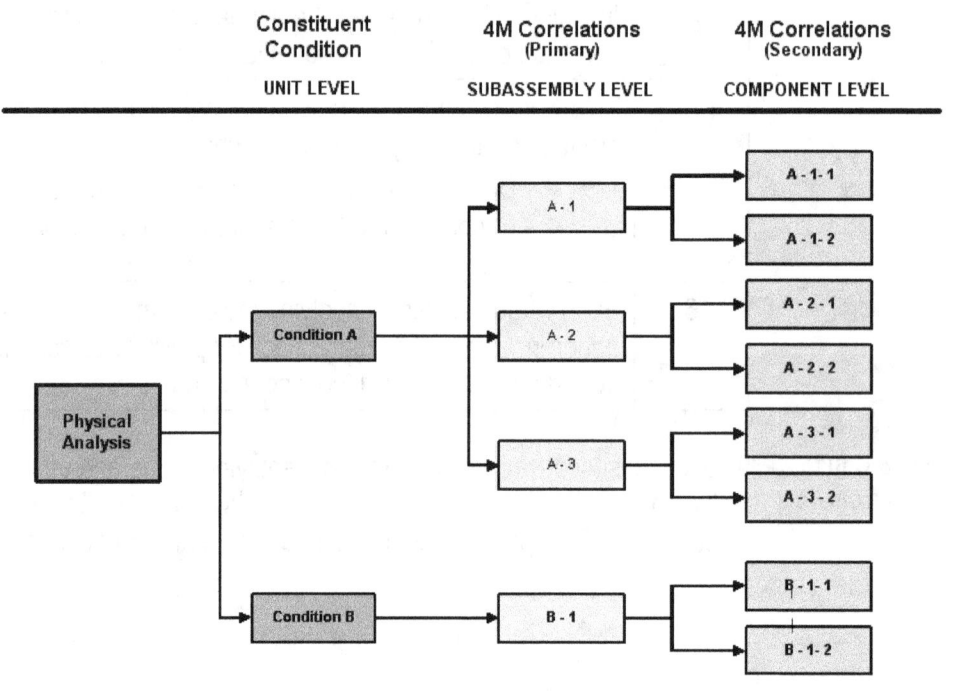

Continued on next page

P–M Analysis Approach, Continued

Relationships

The following example shows the relationships between constituent conditions and production inputs:

Phenomenon:		The bagger shuts down during normal production run.
Constituent Condition:	1.	Product not reaching the bagger
	2.	Film not aligned on feed tube
Primary 4Ms:	1.1	Filler Tube starved of product
	2.1	Film not properly threaded in the sealer unit
Secondary 4Ms:	1.1.1	Scale unit is jammed
	1.1.2	Timing on scale is out of spec.
	2.1.1	Forming shoulder out of alignment
	2.1.2	Transport belt misaligned or loose
	2.1.3	Incorrect threading by operator

The Cause & Effect Chain

The relationships are expressed in terms of cause and effect. In the above example, the secondary 4Ms are the cause of the effect shown as the primary 4M. The constituent condition is the effect of the causal factor shown as the Primary 4M. The constituent condition is the cause of the effect which shows up as the phenomenon This is illustrated in Figure 19.

Continued on next page

P–M Analysis Approach, Continued

Figure 19:
Linkage of the
Three Factors

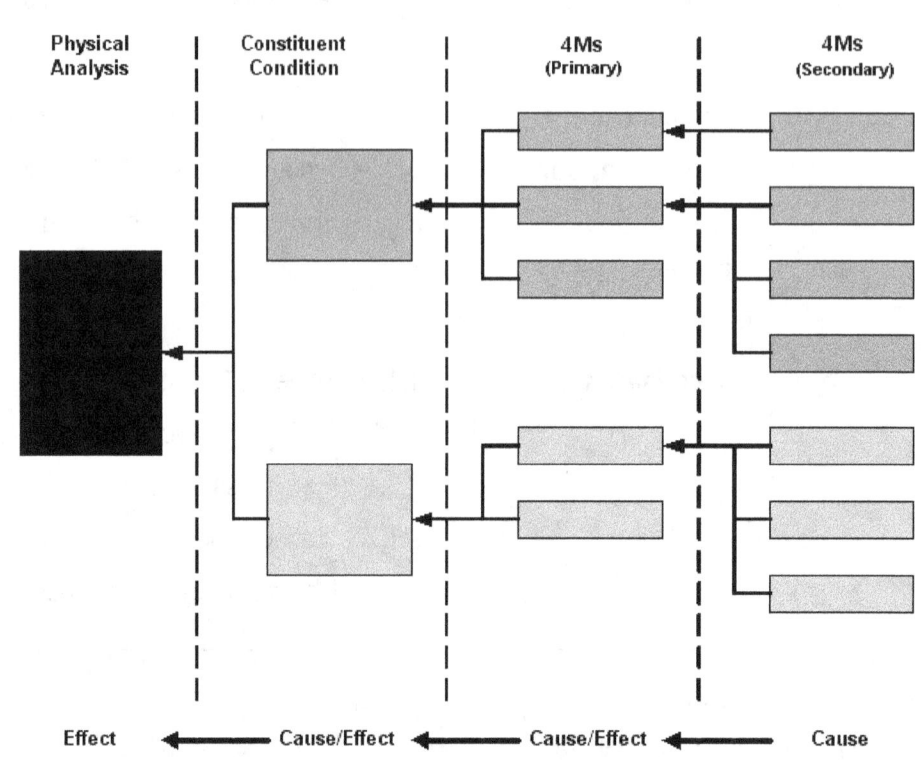

| Physical Analysis | Constituent Condition | 4Ms (Primary) | 4Ms (Secondary) |

Effect ← Cause/Effect ← Cause/Effect ← Cause

Deriving Primary 4M Conditions

It is important for the Reliability Team to consider the following when identifying the Primary 4M Correlations:

- Ignore extent of impact or degree of contribution to the constituent condition. At this stage, prioritization is not considered.
- Assess correlations by working progressively within each subassembly from the surface or top to the supporting base.
- Make sure that you list all possible items for each of the 4 Ms
- Review the resulting primary list to ensure that they actually result in the constituent conditions, remove any from the list that do not.
- Express all 4M elements in measurable or verifiable terms.

Continued on next page

P–M Analysis Approach, Continued

Deriving Secondary 4M Conditions

Take each primary correlation on your primary list and reduce it to its constituent elements at the component level. Determine if it contributes to a corresponding primary factors Just as in the Primary 4M conditions, be sure and express all 4M elements in measurable or verifiable terms.

Move Toward Identifying Solutions

Once the Reliability Team has identified all probable causes, they are ready to move toward identifying solutions. The next steps will involve establishing optimal conditions.

Establishing Optimal Conditions

Overview

Introduction

Up to this point the Reliability Team has defined the phenomenon, analyzed it and determined all of the factors that might be linked to it at three levels:

- Mechanism
- Subassembly
- Component.

In this next section the Reliability Team must search for and identify all abnormalities within these factors

Description of Optimal Conditions

Optimal conditions represent equipment operating at its highest level of reliability. Optimal conditions are the sum of:

- Necessary
- Desirable

Necessary conditions are the minimum necessary to support operation. Desirable conditions are not essential for operation, but they contribute to the prevention of losses such as breakdowns or defects. When these conditions are missing, they are called abnormalities. It is hard to determine if they are missing unless clear standards exist for comparison.

In This Chapter

The topics covered in this chapter include:

Topic	See Page
Basis for Identifying Abnormalities	83
Defining optimal conditions – Standards	85

Basis for Identifying Abnormalities

Introduction

The Reliability Team can not rely on mere subjective judgments; it must establish reliable criteria for determining if each potential cause listed is normal or abnormal. The purpose is to prevent reoccurrence of the problem.

Determine Standards and Criteria

The team starts by surveying the standards and criteria established by the designer of the particular process being analyzed. Often times no suitable criteria or standards exist due to in-plant engineering changes, lost reference material and operations manual or other design documents. In such a case, the Reliability Team will have to develop new ones that properly reflect the operating principles, product quality requirements, and product specifications.

Table 7: Typical 4M Criteria

The following table provides possible sources of 4M related criteria:

4Ms	Standards and Criteria
Equipment Precision	• Determine normal values from equipment documentation and drawings • Review equipment inspection records to determine normal values • Check maintenance records and any PM procedures available for equipment • Establish standards based on equipment function, design, and quality requirements.
Human Factors	• Clearly identify what is required to adhere to all task standards.
Process Quality	• Use quality criteria to identify any characteristics that must be maintained from previous processes.
Level of Standards	• Checkout and understand any and all task standards and procedures for equipment setup, operation, inspection, and shutdown. Verify that those standards and procedures are current and accurate.

Continued on next page

Basis for Identifying Abnormalities, Continued

Determining Boundaries

It is important for the Reliability Team to be vigilant when determining the boundaries between what is normal and abnormal operation. The line of demarcation can be tricky at best as there are a lot of gray areas to be considered. Most teams will find that there is a very high probability that the phenomenon they are considering are linked to defect generation. The team should consider:

- When there is a clear dividing line, shift it in order to narrow the range
- If there is a gray areas, fix it so that it aligns toward the normal side or if that is not possible, narrow the range.

Defining Optimal Conditions - Standards

Introduction

Optimal can be defined as the way things ought to be, not necessarily how they have always been. There are two factors in optimal:
- What is necessary for proper operation?
- What is desirable to ensure proper operation?
-

Another way of saying that is that optimal is a means of ensuring "Zero Defects"

When There Are No Standards

When there are no standards in place or the effectiveness of standards are questionable, the Reliability Team must conduct analyses to establish and confirm them. Typically, the team would set provisional standards and Implement them on a trial-and-error basis.

Establishing Optimal Conditions

Use the following procedure for establishing optimal conditions:

Step	Action
1	Start at the beginning and study all relevant standards.
2	Avoid the temptation to take existing standards at face value; look at how functions and structure relate to quality standards. Study all relevant operations and maintenance manuals and drawings to determine true optimal conditions. Remember to check out any as-built drawings, or changes to equipment.
3	Clarify the boundary between normal and abnormal
4	Stick with it and don't get discouraged when there are no existing standards. Use analysis to develop new ones
5	When working with appropriate standards to determine optimal conditions, continue to seek improvement in operating conditions while maintaining those standards.

Determining Causal Factors

Overview

Introduction	Surveying causal factors in order to determine abnormalities is a difficult and complex job. The Reliability Team needs to: • Determine the most reliable and efficient method for measuring the gap between causal factors and the ideal values identified in the analysis • Determine the best way to survey all of the factors at the equipment location • Survey the factors in such a way that targeted values are properly measured and compared to optimal standards to identify abnormalities.
Planning Best Use of Time	It is important for the Reliability Team to start with the constituent conditions. This allows the team to eliminate those conditions that are found to be free of abnormalities. The team can then eliminate them and their primary and secondary 4M correlations, thereby saving time and expense. This will speed up the improvement process significantly.
In This Chapter	The topics covered in this chapter include:

Topic	See Page
Identifying Probable Cause	87
Root Cause Analysis	88

Identifying Probable Cause

Introduction

Reliability Teams should review all items listed as causal factors where possible. In some cases this is not practical as it will be too time consuming. The team may have to streamline the list as shown below.

Survey Procedure

The survey procedure can be streamlined by taking it one level at a time and using the following procedure:

Step	Action
1	Measure the status of all items, if they are normal, skip the 4M correlations.
2	Review and measure primary 4Ms only for constituent conditions whose measured values are different from the standard values.
3	Review and measure secondary 4Ms only for primary 4Ms whose measured values are different from the standard values

Determine the Abnormalities to be Addressed

Once the team has identified all of the causal factors and surveyed their condition with the proper measurement tools, they can decide which deviating conditions should be considered true abnormalities and what the probable cause is.

Identify an Abnormality

The following procedure should be used by the Reliability Team to identify abnormalities:

Step	Action
1	Thoroughly investigate all factors
2	Compare abnormal conditions against current or provisional standards
3	Think in terms of "optimal" conditions not just necessary conditions.
4	Classify any items on the border between normal and abnormal as abnormal.
5	Ensure that you understand all of the causal factors behind each condition identified as abnormal.

Continued on next page

Root Cause Analysis

Introduction	Root cause determination is the process used by the Reliability Team to systematically detect and analyze the probable causes. During root cause analysis, the team relies on internal logic and reasoning skills as well as research to reach a conclusion. The team makes their thinking visual using various tools and methods. This opens up the process to team based synergy, which strengthens the process.
Purpose	The purpose of root cause analysis is to: • Determine presumptive causes of the problem • Eliminate apparent and presumptive causes that are not supported by data • Select causes that need verification • Determine the root and contributing causes that require corrective action.
Importance of Determining the Root Cause	It is important to distinguish between the primary or root cause and the contributing causes in order to develop the necessary corrective actions to prevent the problem from reoccurring. Without a thorough investigation of the problematic situation, you may initiate corrective action that does not eliminate or alleviate the problem and wastes valuable resources.
Beyond Root Cause	Once you have defined the problem based on facts, you can focus your root cause analysis efforts, plan a strategy, and begin to obtain the data needed to hypothesize and test possible cases. Keep in mind that root cause analysis is not the endpoint for the Reliability Team; it is really the starting point for resolving chronic losses in your equipment. The next step in the process is to apply the CEDAC diagram.

Creating a Reliable Methodology - CEDAC

Overview

Introduction	The CEDAC diagram was established to handle situations in which a reliable method has not been established. This will require the setting of a standard by integrating the knowledge and best practices of the entire team. It may also require the temporary participation in the team by consultants and experts from other areas of the company.
Standards	Earlier in this book we addressed the setting of standards, in the case of setting a reliable method, it requires special attention to preventing the occurrence of defective outcomes. If current standards are allowing defects or problems to arise, they must be changed. The CEDAC Diagram is a useful tool for accomplishing this.
In This Chapter	The topics covered in this chapter include:

Topic	See Page
A Tool For Continuous Systematic Improvement	91
Window Analysis	93
How To Make And Use a CEDAC Diagram	96

A Tool for Continuous Systematic Improvement

Introduction	CEDAC was developed by Dr Ryuji Fukuda as a method for developing effective standard procedures. It is an extremely valuable tool for helping the Reliability Team overcome chronic losses and improve equipment reliability. It works particularly well with PM–Analysis.
Benefits of the CEDAC System	CEDAC accomplishes the following: • Energized improvement activities and promotes steady progress towards desired results • Establishes an atmosphere of "Continuous Improvement" • Changes organizational culture which leads to sustained organizational change • Is a versatile tools that can be used anywhere by anyone regardless of industry or conditions • Communicates progress or lack thereof
Foundation of CEDAC	CEDAC is built upon three driving forces: • Development of reliable systems • Create a favorable environment for improvements • Application of the process in daily work
Logical Choice of Reliability Teams	CEDAC is a logical choice to help the Reliability Team reduce chronic loss and improve Overall Equipment Effectiveness. Accomplishing these goals is a long-term project and the CEDAC system is well suited to assist the team.

Continued on next page

A Tool for Continuous Systematic Improvement, Continued

CEDAC Diagram Components

The CEDAC diagrams consists of two parts as shown in Figure 20:
- Cause Side of Diagram
- Effect Side of Diagram

Figure 20: Typical CEDAC Diagram

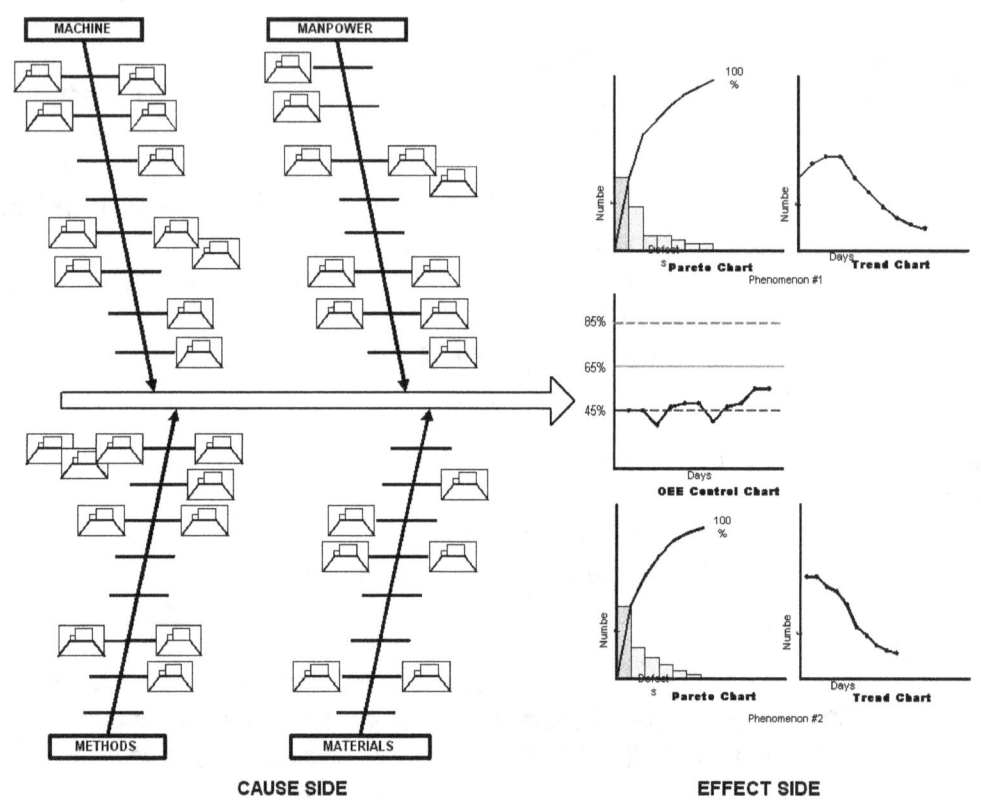

CAUSE SIDE   EFFECT SIDE

Window Analysis

Introduction

The windows analysis is a tool to help the Reliability Team bridge the gap between the PM–Analysis and the CEDAC diagram. It allows the Reliability Team to analyze concrete facts about the various causal factors which may be causing a particular phenomenon to occur in the equipment.

Purpose

The windows analysis technique was created to help the Reliability Team establish habits for gathering facts correctly, categorizing data for measurement, and establishing effective counter measures.

Figure 21: Typical Window Structure

X \ Y		KNOWN		UNKNOWN
		Practiced	Unpracticed	
KNOWN	Practiced	A	B	C
	Unpracticed	B	B	C
UNKNOWN		C	C	D

Continued on next page

Window Analysis, Continued

Windows Analysis Structure

Figure 21 shows the basic structure of a typical window used in windows analysis. Person X and person Y represent two different backgrounds or experience levels within the Reliability Team. Typically, person X would typically represent a manager or senior level person within the team. Person Y would represent a new comer to the team or junior person with less experience.

Table 8: Key Windows Analysis Concepts

The following table defines the four key concepts for windows analysis:

Term	Definition
Known	The standards and procedures for preventing defects are established and communicated to all concerned
Unknown	The standards and procedures for preventing defects are not established yet
Practiced	The procedures are practiced 100% of the time.
Unpracticed	The procedures are not practiced 100% of the time

Information Categories

Figure 21 also shows the four different categories used in windows analysis:
- Category A
- Category B
- Category C
- Category D
-

These categories are explained in Table 9

Continued on next page

Window Analysis, Continued

Table 9: Windows Analysis Categories

The following table explains the four categories used by the Reliability Team when conducting a windows analysis:

Category	Description
A	The correct procedures for preventing defects are established and both parties know and correctly practice the procedures.
B	The correct procedures for preventing defects are established and there is someone who does not practice them correctly. This is an adherence problem. This includes three subcategories 1. Although the correct method is known, there are careless mistakes resulting in non-adherence to the procedure. 2. Although the correct method is known, someone lacks the skill and cannot utilize the knowledge. 3. Although the correct method is known, lack of time, manpower and/or money leads to non adherence to the procedure.
C	The correct procedures for preventing defects are established, but one of the two parties who should have been informed, does not know the procedure. This is a communications problem
D	There is no correct standard or procedure for eliminating defects. Because there is no resolution of the technical problem, neither party knows how to eliminate the problems.

How to Make and Use a CEDAC Diagram

Introductions   CEDAC is an effective system for reducing manufacturing defects and streamlining manufacturing processes. It can be effectively used to identify and resolve very complex problems. The shape of the CEDAC Diagram is similar to the Cause-and-Effect diagram; however the goal is completely different.

CEDAC Diagram Procedure   Use the following procedure for creating and using the CEDAC Diagram:

Step	Action
Pre Diagram Activities	
1	Determine what needs to be improved. Look to results of the PM–Analysis to determine this. Identify how the results should be measured.
2	Select a target for improvement. What are the profit goals as a result of the improvement? How does it impact • Throughput • Inventory • Operating costs?
3	Identify problems and obstacles which might prevent attainment of the target
4	Establish standards to be adhered to in order to achieve good results.
Creating The Diagram	
5	Draw the Diagram: Effect side on the right – the cause side on the left
6	Define the focus of the project
7	Select the project leader
8	Establish project results measurements and link to the effect side.
9	Set the target
10	Format the effect side by creating the appropriate charts and diagrams

Continued on next page

How to Make and Use a CEDAC Diagram, Continued

CEDAC Diagram Procedure **(continued)**

Step	Action
	Using The Diagram
11	Gather the "Fact" cards
12	Generate Improvement cards
13	Test the improvement ideas
14	Choose the Standard Cards
15	Write a Detailed Process Sheet or procedure

Starting Point

The starting point of the CEDAC Diagram begins with the leader of the Reliability Team making the final decisions for the direction of the improvement project. The leader will of course consider the data collected through the analyses and the input from team members and applicable outside parties. The Windows analysis is used as a final filter. The Effect side of the diagram is determined by the leader of the Reliability Team in conjunction with Management.

Continued on next page

How to Make and Use a CEDAC Diagram, Continued

Table 10: Measuring Results

Use the following table to determine measurement of desired results

Improvement	Measurement
Throughput Improvement	• OEE • Amount of work in process • Fraction of defects • Set up time • Changeover time • MTBF • Frequency of failures • Production lead time • Profitability of products
Inventory Reduction	• Average inventory level • Inventory turnover (annual sales/average inventory level) • Amount of raw materials waiting processing
Operating Cost Reduction	• Labor productivity • Number of defects • Failure costs • Machine productivity • Product costs

Continued on next page

How to Make and Use a CEDAC Diagram, Continued

Effect Side Components

The Effect Side of the diagram must be decided by the leaders. Targets for improvement must be closely tied to company strategic plans, goals for profitability, cash-flow and key performance indicators. There are three key charts which must be part of the CEDAC diagram (others may be added if necessary depending on the results to be tracked):

- Raw Data Chart
- Pareto Chart
- Trend Chart

Figure 22: CEDAC Effect Side Example

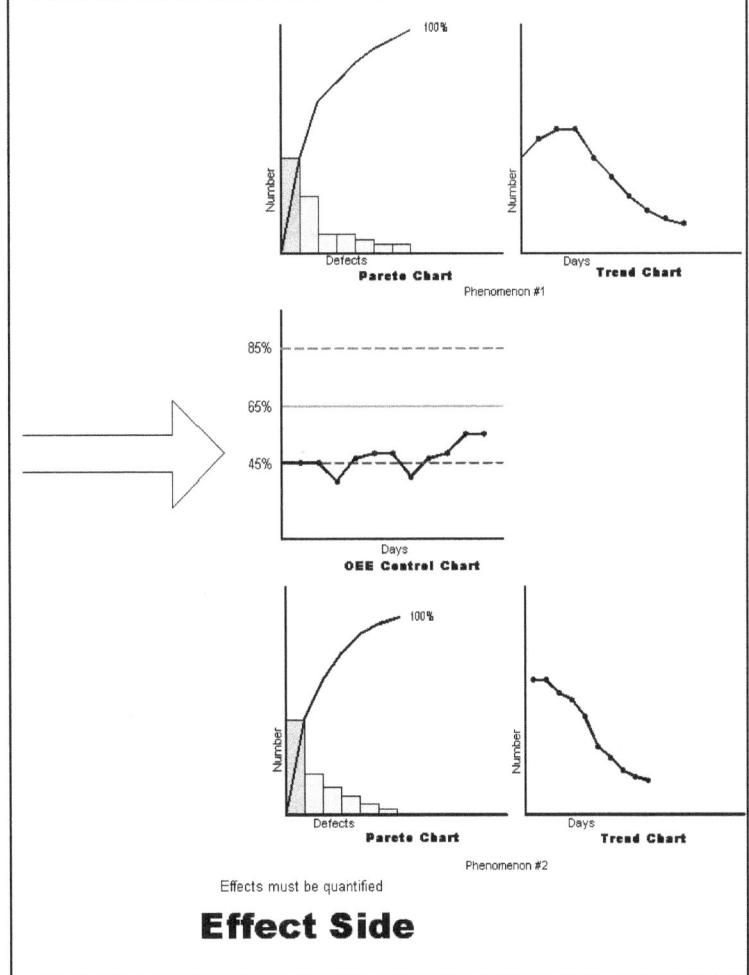

Continued on next page

How to Make and Use a CEDAC Diagram, Continued

Developing the Cause Side of the Diagram

One important feature of the Cause Side of the Diagram is that it involves all members of the Reliability Team equally. The first step in creating the Cause Side is to write all of the possible obstacles to the goal on individual cards. These are known as Fact Cards. They are then categorized into one of the following categories and placed on the left side of the spine:
- Unusable
- Of Interest
- Under Preparation
- Under Test

Figure 23: CEDAC Cause Side Example

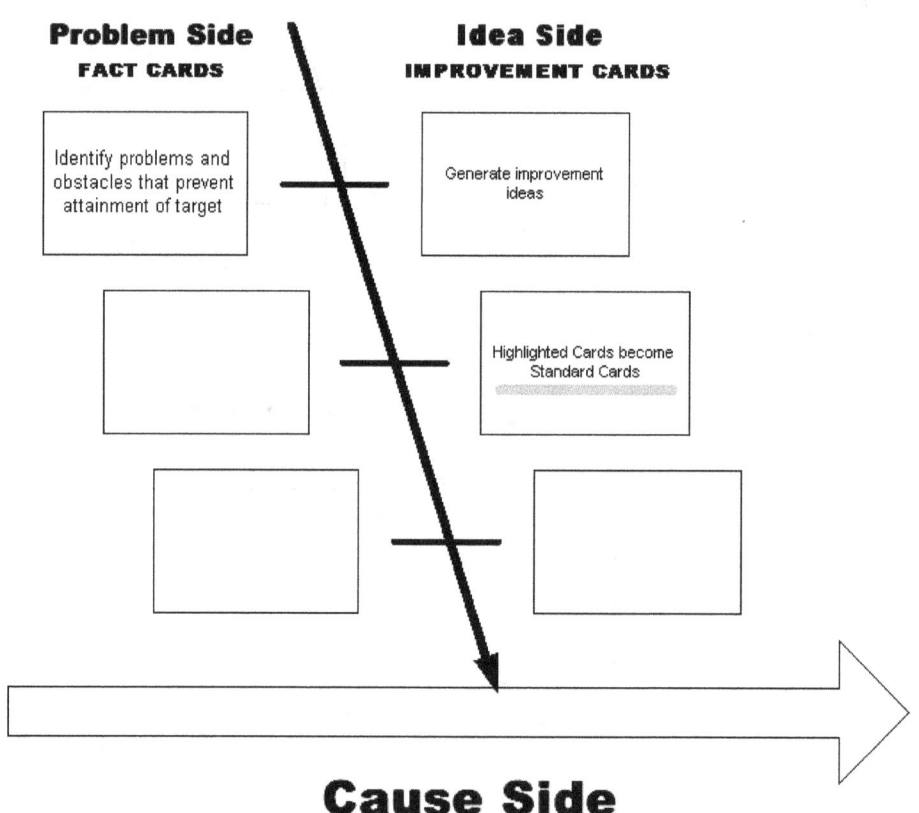

Continued on next page

How to Make and Use a CEDAC Diagram, Continued

Developing Improvement Cards

Once all of the fact cards have been categorized and displayed, the Reliability Team turns its attention to developing "Improvement Cards". The Reliability Team considers the fact cards and develops ideas for improvement. Each improvement idea is tested and the results charted on the effect side of the diagram. The Improvement cards are attached to the right side of the spine.

Features of CEDAC

There are two distinct features of the CEDAC Diagram:
- **The Effect Side** – is directed by the organizations plans, objectives and policies.
- **The Cause Side** – requires complete participation of the team

Standard Cards

Improvements that work are put into practice and the results tracked on the effect side. Successful improvements are highlighted in green on the effects side. The green highlighted improvements become standards. These cards are now known as standard cards as shown in Figure 23.

Detailing the Process

Once the standard card is developed, the Reliability Team writes a written procedure or Detailed Process Sheet to implement the new standard. The new procedure is then communicated to all concerned and must be practiced by all concerned. The results are tracked on the effect side for a period of time. It the desired results are not achieved the standard is reviewed using windows analysis and changed made as required.

Continued on next page

How to Make and Use a CEDAC Diagram, Continued

The Detailed Process Sheet

Figure 24 shows a Detailed Process Sheet from QCD Systems' book: *The Next Step*. This is an excellent tool for detailing the process and incorporating the "Standards Cards" as shown in Figure 23. In *The Next Step*, the author asks four (4) questions that are critical to the success of the Reliability Team:

"*At This point it is important to step back and ask the following four questions:*
1. *Do we have a written process?*
2. *Is that process capable of doing what it was written to do?*
3. *Are those using the process properly trained?*
4. *Are they being regularly audited?*"[1]

Figure 24: Typical Detailed Process Sheet

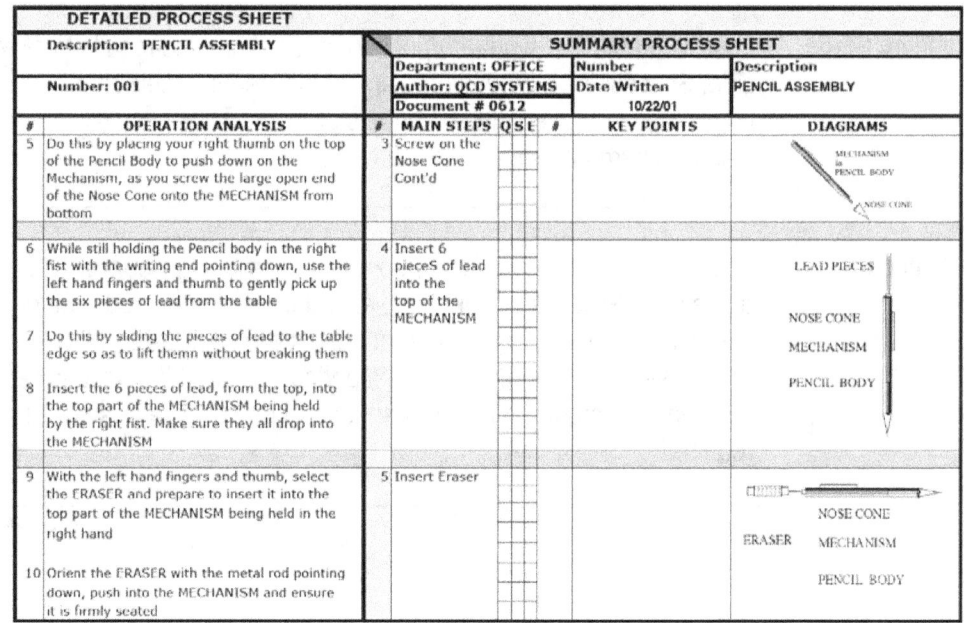

Continued on next page

---

[1] The Next Step, Peter Paola, QCD Press, Chicago, IL , 2004, p 56

How to Make and Use a CEDAC Diagram, Continued

The Answers Must Be YES!

If the Reliability Team is to be successful at reducing chronic losses, all 4 answers to the above questions must be an emphatic YES! The Next Step contains an explanation of how to write a proper Detailed Process Sheet.

I highly recommend that you pick up a copy of The Next Step at the QCD Systems website: www.qcdsm.com. It has a wealth of information that will supplement the information in this book.

Other Programs from ALERA

ALERA BOOKS AND PROGRAMS

Topic	See Page
Books	
Training Programs	
Services	
5S Blog	

ALERA Group Website

http://www.aleragroup.com

Books From ALERA Publishing Group

5S Topics

Planning & Implementing 5S

Paperback	$21.52
Hard Cover	$36.53

The Planning & Implementing 5S program shows you how to organize a Performance Improvement Steering Team, how to analyze the workplace, how to plan a facility-wide improvement program, and how to sustain your efforts.

5S Workbook

Paperback	$14.48

The 5S workbook is the companion to Planning and Implementing 5S by Brice Alvord. It provides the tools used in the ALERA workshop.

Auditing Your 5S Program

Paperback	$12.24

The 5S audit is critical to the success of your 5S program. It is often overlooked or considered an unnecessary extra expense. The audit validates the accountability of the target area owners for complying with 5S plans. Without the audit, the program slowly withers away and becomes ineffective. A close look at 5S failures will reveal a lack of or an ineffective auditing program. This book explains how to conduct a proper 5S audit.

Creating Sound Financial Based 5S Projects

Paperback	$17.99

Creating Sound Financial Based 5S Projects is intended for non-financial managers who are involved with planning and implementing 5S programs. The book provides an overview of financial management as well as a methodology to create a financially sound 5S program that is designed for success. It presents how to establish a project matrix based on proven logic, indicators of effectiveness, and analysis of external factors that impact your project. This book is intended as a follow on to our successful Planning & Implementing 5S book.

Continued on next page

Books From ALERA Publishing Group, Continued

Training

How to Train People On The Job

Paperback $18.53

NEW and REVISED workbook for Training On the Job Trainers. Covers adult learning theory, why shadow training does not work, how to perform a simple job/task analysis, how to develop trainer's guides and teach using the Four Step Method

Performance Based Training

Paperback $21.95

This book is intended as a guide to Performance Based Instructional Design. It covers how to conduct an effective Training Analysis including Job/Task Analysis, how to identify and define realistic competencies and instructional objectives and how to organize analysis data into a performance based training design. The book also explains how to develop important training documents including trainer's guides and lesson plans, participant manuals, and support materials including training and job aids and other media. Performance Based Training: Building a Competent Workforce is intended for the training professional as well as those people who have been given a training assignment. It is also a good reference for managers and supervisors to help them build a stronger workforce and to support company training efforts.

Advanced Instructional Design

Paperback $21.95

Advanced Instructional design focuses on the steps required to develop a performance based training design. Chapters include information conducting a Job Task Analysis and the Design of the training program. Other topics include defining competencies, conducting a DACUM, writing performance based objectives, developing criterion tests, Sequencing training elements, and writing a training blueprint. This book does not cover the development of training materials that will be addressed in another book yet to be published.

Continued on next page

Books From ALERA Publishing Group, Continued

Training, Continued

One Point Lessons

Paperback $11.94

This book is a training workbook for developing One-Point Lessons. It is designed to provide clear and simple explanations of procedures and techniques to quickly create short, cost effective training materials.

Other Topics

Creating A Performance Based Culture In Your Workplace

Paperback $34.99

Creating a Performance Based Culture in your workplace is a nuts and bolts approach to planning and implementing a performance based continuous improvement program for your facility. It shows you how to incorporate strategic planning and business needs analysis into a strong program that addresses your business needs and related performance issues. It shows you how to create a strong business case for change and how to create structured on job training designed to carry out that business case. Filled with illustrations, charts and procedures. Includes state of the art tools to help improve your organization's performance and improve your bottom line.

Overall Equipment Effectiveness

Paperback $17.56

Overall Equipment Effectiveness (OEE) is a universal measurement that has been used worldwide for over 10 years. It is a formula to measure the efficiency of production line equipment. In short, OEE measures the ratio of first-pass acceptable product actually produced to the theoretical amount that could be produced under optimal conditions.

Strategic Thinking

Paperback $13.16

Strategic Thinking for Operations & Projects focuses on how to build a strategic based business case for change. It is a powerful communications tool for getting projects approved.

Continued on next page

107

Books From ALERA Publishing Group, Continued

Other Topics, Continued

Fundamentals of Project Management

Paperback $16.38

Project Management Fundamentals covers the fundamental skills required to plan and implement project. It is intended for new project managers and managers with little or no project management experience.

Training Programs From ALERA Group

How To Train People On The Job

A 2 Day Hands-on Workshop that teaches your participants how to conduct On Job Training using the Four Step Method of Instruction.

Planning & Implementing 5S Workshop

A 3 Day Hands-On Workshop that teaches your participants how to plan and implement a basic 5S program. They will actually begin implementing 5S in a target area of your facility.

Team Based Problem Solving

A 2 Day Hands-on Workshop to teach your teams how to work together to identify and solve real problems in the workplace. Teams will address n actual problem and apply the tools to solve it.

Project Management Workshop

A 2 Day Hands-on Workshop that teaches the fundamentals of project management. Participants develop all of the elements of a project.

Overall Equipment Effectiveness

A 1 Day Hands-on Workshop to teach your participants what Overall Equipment Effectiveness is and how to calculate it accurately.

Try Z Seminar

A 2 ½ to 4 Day Hands-on Workshop from QCDSM Systems, Inc. See QCDSM Program Information on page **Error! Bookmark not defined.**.

Services From ALERA Group

Introduction	The ALERA Consulting Group exists to assist you in improving all areas of performance in your organization. We have a variety of state of the art tools and processes to help you identify performance needs and relate them to business practices and strategies.
Strategic Thinking	ALERA helps you develop strategic thinking in your organization; We conduct a strategic thinking workshop for selected members of your management team. We coach them through the application of the Strategic Thinking model to help them develop comprehensive and effective business cases for change within your organization.
Team Building	ALERA helps you design and deliver the right customized team development program, team building event, corporate retreat, or executive retreat that will improve your team's effectiveness, collaboration skills, and team-based results.
New Leader Assimilation	ALERA's Leader Assimilation program is based on the process designed by Kaiser Aluminum. Kaiser discovered that it normally took an incoming manager six months to become fully productive. The process was designed to reduce this amount of organizational down-time.
High Impact Change Management	ALERA's High Impact Change Management is a 3 phase organizational design program that assesses the organization with a performance audit, rationalizes change, and develops a comprehensive design. ALERA provides: design team formation and training, strategy focused design, alignment of the organization for maximum effectiveness, and building an empowered organization.
Asset Effectiveness (Focused Equipment Improvement)	ALERA helps your Operations Improvement Team develop the skills to address chronic equipment problems that hinder your profitability and overall performance. We provide workshops o help your team succeed. We evaluate your team's performance and coach individual members and your management team.

Continued on next page

Services From ALERA Group, Continued

Training Analysis	ALERA conducts or teaches your team to conduct a variety of training analysis including: training needs analysis, job/task analysis, cost benefit analysis, Our training professionals conduct training effectiveness audits, subject matter interviews, and individual performance evaluations.
Workplace Organization (5S planning and implementation)	ALERA helps you analyze your needs, design a program, plan 5S implementation, evaluate the progress of your program, perform 5S audits, coach your management team on implementation problems and opportunities.
Project Management	ALERA has experienced project managers who can assist you with keeping your performance improvement project or training project on schedule and under budget. We address your organizational needs and support consulting efforts with comprehensive training programs for your team as needed.
Technical Writing and Instructional Design and Development	ALERA can provide you with technical writers to help you develop standard operating procedures, lockout/tagout procedures technical documentation, training manuals, detailed process sheets.

Continued on next page

Services From ALERA Group, Continued

The following program is offered by QCD Systems:

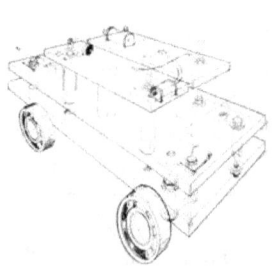

Try–Z Workshop

2 ½ –Day Workshop

The Try-Z training session is uniquely designed to take managers and executives through an introspective 2-1/2 days of exploring and understanding the challenges involved in creating a continuous improvement environment.

Upon completion of the Try-Z seminar will be able to create the environment that will allow them to solve the organization's specific problems by applying QCDSM principles.

Prerequisite: None

For More Information:

PHONE
 (610) 927-0390

EMAIL
 info@aleragroup.com

FAX
 (610) 927-0916

Excerpts from *"The Next Step"*

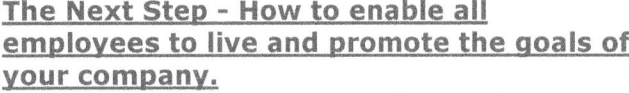

The Next Step - How to enable all employees to live and promote the goals of your company.

COLOR EDITION *(Preferred for graphics contained in the book)*

The Next Step enables management to encourage employees to own goals of the company by enabling them to manage their own areas and track tasks through continuous improvement. The two key elements of this methodology are the development of authentic procedures by the employees themselves and the daily measurement of achievement against these procedures. The ownership results in a self-driven desire to meet targets and overcome daily issues.

From: $29.95
Black & White From: $14.95

The Next Step

Excerpts From "The Next Step"

Excerpts From "The Next Step"

Excerpts From "The Next Step"

INTRODUCTION

In the competitive world of business, any company must ensure that they are doing what is necessary to not only maintain their market position, but also make sure that they are keeping abreast of the latest manufacturing and service related improvements.

OEE – Overall Equipment Effectiveness is an important tool in maintaining efficiencies and productivity. The main thrust of OEE is to obtain maximum availability performance and Quality from the equipment that is used in your facility. Therefore there are two important factors to be taken into account in this regard. Firstly, ensuring that your equipment is functioning at its maximum but secondly, that you also have in place a methodology that enables your people – they work the equipment – to be directly involved in maximizing your effort.

WHY DO YOU NEED A NEXT STEP?

I am sure that there are many of you who have asked yourselves how you can successfully transfer your vision for your company or your group or what ever it is you are leading in such a way that the people are as enthusiastic and committed as you are?

I have visited so many companies where there are diverse 'agendas' being followed. Any attempt to introduce your vision or the ubiquitous 'mission statement' to the employees as a whole and to managers in particular seems to fail. The marshalling of resources and the buy in needed to steer the ship along one path seems to take up so much time with little result.

I believe that I have a mechanism, a process, which facilitates this and delivers remarkable results in so many other different ways as well.

I call it the QCDSM process and any person or organization that wishes to move forward and benefit from new technologies and developments etc. can use this process to reach those goals.

I will outline for you the philosophy behind the process or system and then go into some detail as to how it works.

There is no substitute, however, for a hands on, one on one conversation about the process so I invite you to contact me at qcdpjp@ix.netcom.com or our web site: www.qcdsm.com

THE BENEFIT

There is a need in business and even in one's personal life as well, to go on a retreat! By this I mean to give yourself the opportunity of taking the time to regroup and to reassess where you are going and whether you have the means to achieve your goals.

Companies also need to go on this 'retreat' and should do so fairly often. The market place is changing very rapidly and new technologies can catapult your competitors ahead of you. Without really retreating to move forward, major mistakes can be made.

After your 'retreat' and a clear assessment of where the company stands, only then can mission statements other matters be dealt with.

But, so often companies fall into the trap of believing that by establishing goals and by developing mission statements that this will facilitate the move forward without really retreating and asking difficult questions about their present status. Meetings are held to promote the goals. Off site 'conventions' are conducted and great effort is put into announcing the 'mission statement', the goals and the way everyone should follow.

Believing your own advertisements then becomes the greater danger and while all over the facility there are banners and statements and 'rah rah' type announcements, business seem to just go on as usual without much change.

You have committed the most fundamental mistake! You have not developed a process on how to achieve the goals you really want to reach and as a result everyone interprets 'your' goals differently.

The secret is to entrench your goals within a process that will lead the company and the people to the desired finish.

THE PROCESS

No matter what it is you want to achieve, the only way of doing it after the goals have been clearly developed, is to implement a process that will lead the company, step by step, along the desired path.

I call this process the QCDSM System and it is designed to install a structure and a discipline throughout the company so that all can focus on whatever goals need to be achieved. It is a company wide system with checks and balances and with accountability so that from the shop floor, to the laboratory, to the finance depart and human resources, all are focused on achieving the same goals.

THE HUMAN FACTOR

Whatever goals are set, they can only be achieved with the direct involvement of ALL the people of the company. This is true for a sports team, the army and any other organization.

The fallacy is that once the goal has been set, that it will be achieved because the 'managers' will make it happen. Forgetting the vital part that people play in achieving goals is the blunder that has caused the downfall of so many ventures.

If the people of your company are going to be the ones who are going to help you achieve your goals, then you must understand that they need to know **HOW** they have to go about doing this! Very few people are so self motivated as to not only understand your goals or those of the company, but that are also enthusiastic to set about achieving them. In most cases they do not have the 'power' to make the difference!

But, if this is built into their daily work day and if they are given a road map to follow, then the goals and the means of achieving them will be understood. Once understood, their interest will be awakened. They will then be empowered!

Even this is not enough, however. Peaking their interest is but a small step. Getting them involved is the main success. How this is done is what QCDSM is all about.

CHAPTER ONE - WHERE DO WE START

A smorgasbord is where many different types of food are spread out for people to sample and select as part of their meal.

This is what is available in the world of consulting. There is a smorgasbord of training opportunities,. Many and diverse programs and systems are available. Companies and organizations can simply look through this availability and select what they believe fits in with their needs. This will always be the case.

However, the main problem about this approach is that companies fail to realize that inserting 'systems' or training opportunities without focusing on the culture in their company, will not get the desired results.

I have always been of the opinion that a company, no matter how small or big, develops it's own 'culture'. By this I mean their way of doing business. Examples of this can be the way they conduct meetings and conduct business, their presumptions and presuppositions and the opinions they form and develop about their colleagues. 'This is the way we do this here!' could be an expression of this.

Comparing this company 'culture' to what the human body does with its immune system is a good way to describe how the resistance within the company materializes when trying to insert training initiatives. The company 'culture' resists the change in a similar way that the human body rejects 'foreign transplants!'

To overcome this it is necessary to establish a defined procedure to effect the cultural changes within a company if the training initiative is to be accepted. The best way of doing this is to ensure that the structure of the company and the discipline within the company is set up to promote the training initiative and help it become the accepted way of doing business. This also means that the people of the company must be directly involved as well.

QCDSM as a system does this from the very beginning. It is so designed to ensure that from the outset, a structure and discipline is 'inserted' into the culture of the company which not only involves all the people, but which then also ensures that by following the structural design and principles of the process, the journey will be adhered to.

Following this structural process on a daily basis the result will be the development of a new cultural approach in the way in which business will be done to achieve the desired goals. It is not left to individual managers or areas to be the champions or promoters of the system as they interpret it or which allows them to dig in their heels. The task of departments and of individual managers will be to ensure that the structure and discipline is being followed. The structure and discipline is designed to promote the goals.

UNDERSTANDING THE IDEA – AN HOLISTIC PROCESS

Building on this idea, then, QCDSM will examine your organizational structure and determine whether it is set up to accept cultural change in terms of the way in which business is being done. By adapting and selectively enhancing the structure, QCDSM

will develop a customized structure whereby all people within the company will be involved in the implementation of the system to ensure it's success.

Excerpts From "The Next Step"

QCDSM is an holistic process! By holistic is meant, on the one hand, that it cannot be left to individual departments or areas or sections of the company to effect change. It must be implemented throughout the whole company focusing on the goals that will ensure the change that is required. In the other hand, holistic also means that the whole system must be adopted. Many other training initiatives are single, one time interventions. QCDSM is spread throughout the whole company!

Holistic also means that one cannot 'cherry pick' what one wants out of the system. The system as whole is what produces the results.

At this point you may be wondering why it is so important to install a structure and discipline to achieve the goals of the company. The simply answer is that until and unless everyone in the company [holistic] is focused on achieving their key performance indicators on a daily or weekly basis, any attempt to change the culture will be thwarted. The structure and discipline is the process whereby the goals are measured, defined and focused on.

Another important requirement for any company wishing to improve the way it is doing business is to understand, as I have stated previously, the impact of the human factor involved. People will follow you if they understand what it is you want them to do. But, they are unable to follow you if they do not know what it is you want them to do. Therefore, success will be achieved if they are involved in understanding the process of what it is you want them to do.

Finally, there must be a commitment from the top that this is way that the company will be doing business from now onwards. It will require a firm commitment and involvement of not only the top management but every other manager or supervisor or leader within the company. Without this commitment and the championing of the process, it will fail.

Finally, I would like to demonstrate with a graphic the full meaning of the term, 'holistic?"

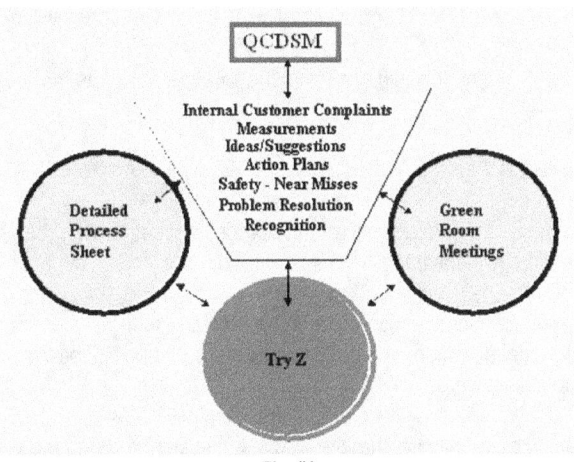

Fig. #1

The three main components of the QCDSM process are the **Try Z Seminar**, the **Green Room Meeting** structure and the **Detailed Process Sheet**. All will be explained in greater detail in the next chapters.

Linking these three main components are various activities and measurements. It is through these

activities and measurement that that the process of managing each area to achieve their goals is obtained.

All are linked together as a unity meaning that each component, while it can exist by itself, nevertheless obtains it's maximum benefit by being linked to the other components. In this way the whole company is involved in the process of continuous improvement.

WHO NEEDS TO BE INVOLVED?

Everyone needs to be involved! After all the purpose of establishing goals and developing mission statements is to move the company forward and to be successful. Everyone must be part of this movement.

The process of getting everyone involved is the greatest challenge. Of course the presentation to the people, the banners, the explanations etc. are all very necessary. The QCDSM process presumes that this will be done at some time.

However, what the QCDSM System establishes within the company is the process whereby the goals outlined and presented are given a tangible, concrete format. Every day or whenever the group or department meet as part of the QCDSM structure, the goals of the area or department and those of the company are measured and discussed and problem solved where necessary.

This is how you get everyone involved. Each day, through the Green Room structure, they are 'talking' one language, as it were: 'Did we achieve our targets for to-day, if not, why not and what are we going to do to rectify the situation?'

WHAT ABOUT THE PRESENT CULTURE VS THE INTENDED CULTURE

Of course this is the greatest challenge and will require firmness. Asking people to change is perhaps the most frightening event in their lives. Everything becomes a challenge and because they feel they are losing their 'roots' or foundation, resistance is great.

Winning them over will be your task but this task can be enhanced and successful provided you explain to them what it is you are asking of them. This is why change is necessary and important? This is the key role they will play in the change.

QCDSM is an employee driven process. Once this is understood the 'change' is not as threatening and in fact becomes a challenge that they will rise to meet.

WHO NEEDS TO TAKE RESPONSIBILITY

Once a decision is taken by the top team to move forward with the process, together with their firm commitment, each manager and supervisor will need to become directly involved.

A QCDSM Coordinator is always selected whose responsibility is not to manage the system, but rather to **coordinate the management of the system** through the managers and supervisors. This person will be the resource person, the coach, the moderator and the 'keeper of the keys' as it were.

Do not forget that the people of the company will become involved in the system through the Green Room process. But, as in all things, this will only be achieved through the leadership demonstrated by the commitment of the management that QCDSM is the way in which business will be conducted!

Excerpts From "The Next Step"

THE FIRST STEPS

Commissioning a training intervention is a very big step in any organization! A decision of this magnitude must be made after considering all the facts and with the concurrence of your major players.

An analysis of what is needed [retreat] can arrive at different conclusions. One example of a conclusion is that a training intervention in one area would solve a problem in that area and that is all that is needed. This is fine.

However, if the greater question is asked about how overall productivity and efficiency can be achieved throughout the company without increasing cost, in fact reducing costs, with the intent to eliminating unnecessary waste, overtime, rework, customer complaints etc., then a whole new scenario enters.

This is a daunting task. Viewing all the training opportunities available, it may take a number of different training interventions to achieve the result unless you can find a process or system that will incorporate all the above requirements into one, company wide intervention that is cost effective and manageable. After all no company can afford to be turned upside down while trying to move forward to the next step.

QCDSM is the holistic system you need. It becomes part of the culture of the company while promoting that culture and incorporates in it's process productivity improvements, efficiency improvements, waste reduction, overtime reduction, rework reduction, continuous improvement processes and many other benefits. It becomes your blueprint for employee participation.

As will be seen in the following pages, the process focuses the company on achieving the goals it has set for itself. Using the talents of the human resources of the company, it channels the efforts of all into the task of achieving the goals and it does this in a very simple way – giving the responsibility for achieving these goals to the people concerned in each area.

PREPARING THE ENVIRONMENT

Like any intervention, the environment in the company must be prepared to receive this training. QCD Systems will work not only with the top team, but with designated leaders to prepare the company for embarking on the training. There are certain requirements that must be established before actual training can begin. Among these are the following:

a. Once commitment is made, both parties must establish the ground rules for the training. This will involve preparing and finalizing the administrative details like contract, payment and other details in this area.

b. The selecting of a QCDSM Coordinator through which QCD Systems will be able to schedule training, coach and mentor the process as it is being implemented.

c. Reviewing the organization structure of the company to ensure that the least disruption takes place as the training proceeds and to ensure that the structure of the company is able to absorb the process.

d. Establish a roll out process that will provide all with the overview of what will be done through the training and the responsibilities that need to be allocate once the various training interventions are accomplished. QCDSM has developed a very detailed structure for this and will share this with each company. It will be customized to suit your company's needs.

e. A decision to take the first step is then made.

www.ingramcontent.com/pod-product-compliance
Lightning Source LLC
Chambersburg PA
CBHW081238180526
45171CB00005B/465